Interdisciplinary Biotechnological Advances

This series is an authoritative resource that explains recent advances and emerging directions in biotechnology, reflecting the forefront of research clearly and reliably, without excessive hype. Each volume is written by authors with excellent reputations and acknowledged expertise in the topic under discussion. The volumes span the entire field from an interdisciplinary perspective, covering everything from biotechnology principles and methods to applications in areas including genetic engineering, transgenic plants and animals, environmental problems, genomics, proteomics, diagnosis of disease, gene therapy, and biomedicine. The significance of these applications for the achievement of UN Sustainable Development Goals is highlighted. The series will be highly relevant for Master's and PhD students in Biotechnology, Nanochemistry, Biochemical Engineering, and Microbiology, medical students, academic and industrial researchers, agricultural scientists, farmers, clinicians, industry personnel, and entrepreneurs.

N. S. Pradeep • Lekshmi K. Edison
Editors

Microbial Beta Glucanases

Molecular Structure, Functions and Applications

Springer

Editors
N. S. Pradeep
KSCSTE-Malabar Botanical Garden
and Institute for Plant Sciences
Kozhikode, Kerala, India

Lekshmi K. Edison
Rajiv Gandhi Centre for Biotechnology (RGCB)
Thiruvananthapuram, Kerala, India

ISSN 2730-7069 ISSN 2730-7077 (electronic)
Interdisciplinary Biotechnological Advances
ISBN 978-981-19-6465-7 ISBN 978-981-19-6466-4 (eBook)
https://doi.org/10.1007/978-981-19-6466-4

This Springer imprint is published by the registered company Springer Nature Singapore Pte Ltd.
The registered company address is: 152 Beach Road, #21-01/04 Gateway East, Singapore 189721, Singapore

Preface

Enzymes are naturally evolved biocatalysts designed with precise suitable regulatory mechanisms which help the cells to survive in extreme environmental conditions by catalysing mostly all the chemical and biochemical reactions. Currently, more than 3000 diverse enzymes have been discovered but only very less is commercially exploited. Microbes are the most preferred source for industrial enzymes because of the easy obtainability and rapid growth rate. The global enzyme market has been rapidly growing over the recent years on account of intensifying awareness and dissemination of industrial enzymes in numerous industrial sectors.

Biotechnology has a direct influence on the industrial sectors, such as food, feed, healthcare, environment and energy, which are mainly focused on socio-environmental essentials, especially economic forces. The understanding of applications in disciplines such as microbiology, biochemistry, bioprocess engineering, genetics and molecular biology merging with the dependency in low energy consumption, utilization of renewable feed stock and environmentally favourable processing potentially led to improved industrial production along with sustainable development.

Microorganisms are the most preferred sources for industrial enzymes because of easy obtainability, and rapid growth rate. Microorganisms have been utilized since ancient human civilization as early as 6000 BC. The prospects of the use of enzymes from microbial sources in industrial sectors have increased significantly in the twenty-first century and endlessly increasing to encounter the demand of promptly growing population and cope with the exhaustion of natural resources. Beta-glucanase are group of carbohydrate enzymes that hydrolyse glycosidic bonds within the insoluble β-glucan substrates, which are unable to be digested by animals easily. These enzymes are characterized extensively in microorganisms like bacteria, fungi and also in plants. They have appreciable ecological implication and commercial applications in established biotechnological industries, viz. brewing and wine production, textile industry, animal feed industry, waste treatments, coffee processing, biofuel production and many more.

This volume contains 12 chapters covering topics related to beta-glucanases. Though the focus is on microbial sources, it also describes other sources of beta-glucanase and its applications. We do believe that this book can provide information to various sectors, viz. academia, industry and anyone who has the interest to handle these topics either as a teacher or as a researcher.

We are thankful to the authors who contributed to this volume. Our sincere thanks to all those who helped us in editing and inspired us in our way.

Kerala, India

N. S. Pradeep
Lekshmi K. Edison

Contents

Editors and Contributors

About the Editors

N. S. Pradeep is currently working as a principal scientist at KSCSTE-Malabar Botanical Garden and the Institute for Plant Sciences, Kozhikode, Kerala, India. He obtained his Ph.D. from Gandhigram University, Madurai, India. He has more than 25 years of research experience in various fields of plant science. Dr. Pradeep has published several research papers in national and international journals and has one patent to his credit. Twelve students obtained Ph.D. under his mentorship so far. He is the author/editor of several books and life member of the Association of Microbiologists of India, Mushroom Society of India, the Indian Mycological Society, the Indian Science Congress Association and the Kerala Academy of Sciences, among others. He is also a fellow of the Kerala Academy of Sciences. Dr. Pradeep is also associated with the Kerala State Biodiversity Board in various capacities. His major area of interest is biodiversity and biotechnology. Dr. Pradeep is also actively involved in science popularization and training, and he was a research committee member of the Integrated Rural Technology Centre, Kerala, India. Currently, he is the governing council member of the Sophisticated Test and Instrumentation Centre (STIC), Kochi, and task force member of the women's biotechnology programs, Dept. of Biotechnology, Govt. of India.

Lekshmi K. Edison was born in Kerala, India. She received her undergraduate degree in zoology and master's degree in biotechnology from Mahatma Gandhi University, Kottayam, Kerala, India, and postgraduate diploma in bioinformatics from Bharathiar University, Coimbatore, Tamil Nadu, India. She obtained her doctoral degree in biotechnology from the University of Kerala in 2019 under the supervision of Dr. N. S. Pradeep at KSCSTE-JNTBGRI, Palode, Trivandrum. The focus of her thesis was structural characterization and heterologous production of beta-glucanases from Streptomyces species. She then secured the research associate fellowship from the Department of Biotechnology, Government of India, and started postdoctoral research at Rajiv Gandhi Centre for Biotechnology (RGCB),

Trivandrum, with Dr. R. Ajay Kumar. Currently, she is working as a postdoctoral associate at the University of Florida, Gainesville, Florida, USA.

Contributors

S. Anu CEPCI Laboratory and Research Institute, Kollam, Kerala, India

K. Anu Chinmaya Arts and Science College for Women, Kannur, Kerala, India

Vipin Mohan Dan School of Biochemistry and Immunology, Trinity College Dublin, Dublin, Ireland

Lekshmi K. Edison Rajiv Gandhi Centre for Biotechnology (RGCB), Thiruvananthapuram, Kerala, India

Yachana Jha N. V. Patel College of Pure & Applied Sciences, CVM University, Anand, Gujarat, India

Shweta Kulshreshtha Amity Institue of Biotechnology, Amity University Rajasthan, Jaipur, India

Karthik Menon Membrane Biology Laboratory, Crystallography and Molecular Biology, Saha Institute for Nuclear Physics, Kolkata, India

T. G. Nandu SpiceHealth COVID-19 RT-PCR Mobile Laboratory, Kasaragod, Kerala, India

P. Nisha Department of Biosciences, MES College, Marampally, Ernakulam, Kerala, India

N. S. Pradeep KSCSTE-Malabar Botanical Garden and Institute for Plant Sciences, Kozhikode, Kerala, India

V. M. Ragitha KSCSTE-Jawaharlal Nehru Tropical Botanic Garden and Research Institute, Thiruvananthapuram, Kerala, India

S. R. Reji KSCSTE-Jawaharlal Nehru Tropical Botanic Garden and Research Institute, Thiruvananthapuram, Kerala, India

P. K. Satheeshkumar Centre of Advanced Study in Botany, Institute of Science, Banaras Hindu University, Varanasi, India

Vinay Sharma Amity Institue of Biotechnology, Amity University Rajasthan, Jaipur, India

Chapter 1
Beta-Glucanases: An Introduction, Marketing Dynamics and Industrial Applications

Lekshmi K. Edison and N. S. Pradeep

Abstract Beta-glucanases are economically important enzymes, believed to have appreciable ecological implications and have commercial usage in several industrial sectors such as food, feed, healthcare, environment and energy, which are mainly focussed on socio-environmental essentials moreover economic forces. Microorganisms are the most preferred sources for industrial beta-glucanases because of their easy obtainability and rapid growth rate. The prospects of the practice of enzymes from microbial sources in various industrial sectors have expanded significantly in the twenty-first century and endlessly enlarging to encounter the demand of a promptly rising population and cope with the exhaustion of natural resources. The microbial beta-glucanase has gained the attention of the universe for its extensive uses in industrial sectors viz. food, medicine, agriculture, chemicals and energy. Due to the high-level performances of beta-glucanase from diverse microbes, which works well under diverse physical and chemical conditions, the production of these enzymes in the industrial sectors is necessitating. Nowadays available beta-glucanases in the market certainly do not encounter the industry's necessities, predominantly due to economic reasons. However, the advancements in genetic engineering and protein engineering can easily be manipulated and cultivated in the microbial cells in large quantities for elevated enzyme production to meet the increased industrial demand. Microbial beta-glucanase usage in numerous industrial sectors is abundant and growing quickly over conventional systems due to being harmless to the environment, better productivity and the good quality products.

Keywords Carbohydrase · Beta-glucan · Global enzyme market · Synergic action

---

L. K. Edison
Rajiv Gandhi Centre for Biotechnology (RGCB), Thiruvananthapuram, Kerala, India

N. S. Pradeep (✉)
KSCSTE-Malabar Botanical Garden and Institute for Plant Sciences, Kozhikode, Kerala, India

N. S. Pradeep, L. K. Edison (eds.), *Microbial Beta Glucanases*, Interdisciplinary Biotechnological Advances, https://doi.org/10.1007/978-981-19-6466-4_1

1.1 Introduction

Biotechnology has a direct influence on the industrial sectors such as food, feed, healthcare, environment and energy, which are mainly focussed on socio-environmental essentials moreover economic forces. The understanding of applications in disciplines such as microbiology, biochemistry, bioprocess engineering, genetics and molecular biology merging with the dependency in low energy consumption, utilization of renewable feedstock and environmentally favourable processing potentially led to improved industrial production along with sustainable development. However, the use of biological processes involving whole cells and cellular constitutes in different industrial sectors emerged with the developments in engineering sciences has been a real advancement, which subsequently enlightens the path to green technology.

Enzymes are naturally evolved biocatalysts designed with a precise suitable regulatory mechanism that helps to survive the cells in extreme environmental conditions by catalysing mostly all the chemical and biochemical reactions. They are the large macromolecules of amino acid polymers, which vary from kilodalton to megadalton in molecular weight. The specificity of substrates is determined by their catalytic sites which are deeply buried within hydrophobic pockets (Singh et al. 2016). With the advancements in biochemistry, a huge number of enzymes were isolated, screened, characterized and finally classified based on the specificity to chemical compounds. The International Union of Biochemistry and Molecular Biology (IUBMB) in collaboration with the International Union for Pure and Applied Chemistry (IUPAC) recognized an Enzyme Commission (E.C.) which guides the enzymes nomenclature and systematic classification (Liese et al. 2006). Currently, more than 3000 diverse enzymes have been discovered; however, only 150 to 170 are commercially exploited. At present only 5% of chemical products are manufactured in industries by using biotechnological aspects. The application of enzymes in industrial processes speeds up the reactions in an eco-friendly and cost-effective manner (Pandey et al. 2010). Enzymes are highly unique in activity and function; hence they can easily be deactivated when needed no longer. It also can execute highly distinct, complex, biocatalytic reactions without any harsh chemistries or utilizing excessive energy makes enzymes more attractive for industrial purposes.

Enzymes are applied to transform or manipulate biopolymers for improving the shelf-life, nutritional properties and for the simplification of nutrient extraction of edible parts. Currently, above 80% of available enzymes in the global market is targeted for the food and feed enzyme industry (Poulsen and Bucholz 2003). In industry, enzymes are applied as the main catalyst for transforming the raw material to the main product; or as an additive to change the functional properties of the product; or as a regulator to regulating the reaction process (Illanes 2008). The driving factors associated to motivate the practice of enzymes from microbial sources in various industrial applications are the collective request of global consumers, the requirement of reduction of price, natural resources diminution and environmental safety (Choi et al. 2015).

Microbes are the most ideal sources for industrially important enzymes because of the effortless obtainability and prompt growth rate. It has been exploited since ancient human civilization as early as 6000 BC. The prospects of the routine of enzymes from microbial sources in industrial sectors have significantly improved in the twenty-first century and endlessly increasing to encounter the demand of a promptly rising population and cope with the depletion of natural resources. The microbial enzymes have gained the attention of the universe for their extensive uses in industrial sectors viz. food, medicine, agriculture, chemicals and energy. In virtue of the superior quality and performances of enzymes from diverse microorganisms, which works satisfactory under diverse physical and chemical conditions, the production of enzymes from microbial in the industrial sectors is necessitated. Furthermore, the advancements in genetic engineering and protein engineering technologies can easily be manipulated and cultivated in the microbial cells in large quantities for elevated enzyme production to encounter the high industrial demand (Liu et al. 2012). Microbial enzyme usage in numerous industries such as pharmaceutical, food, textile, leather, paper, etc. is abundant and growing rapidly over conventional methodologies by reason of harmless to the environment, better effectiveness and good value products (Gurung et al. 2013).

The worldwide enzyme market has been rapidly growing over the recent years on account of intensifying awareness and dissemination of industrial enzymes in innumerable industrial sectors. Globally, the progression of the industrial enzymes market is mainly driven by increasing food and beverage consumption predominantly in meat and wine. Market estimates for industrial enzymes are a sensitive index of biocatalytic processes expansion for commercial-scale production. The industrial market for relevant enzymes applicable in food and animal feed industries in 2021 is estimated to raise globally at an annual growth rate of 4.7% (Chapman et al. 2018). As per Business Communication Company (BCC) Research, the worldwide enzyme market is expected to raise from \$5.01 billion in 2016 to \$6.32 billion in 2021 (Dewan 2018) and \$8.7 billion by 2026 with a compound annual growth rate (CAGR) of 6.3% (Bcc research 2021).

In the global market of industrial enzymes, North America has the largest share followed by Europe and the Asia Pacific. In future also, North America is anticipated to maintain its dominance. Though, the Asia Pacific region is predicted to achieve the highest growth. India and China are expected as main contributors to boost the Asian industrial enzyme emerging markets by investing in multinational food and beverage companies. Brazil and Middle East countries are also expected to boost the global market. While significant growth is expected in the overall value of industrial enzymes in all parts of the world, it leads to an emergence of small to medium scale enzyme industries, development of novel technologies, better enzyme production, increasing demand and rising environmental safety. Advancements in biotechnology and enzyme engineering is a driving force of the global enzyme market by introducing novel products. The cumulative disposable income of customers can afford

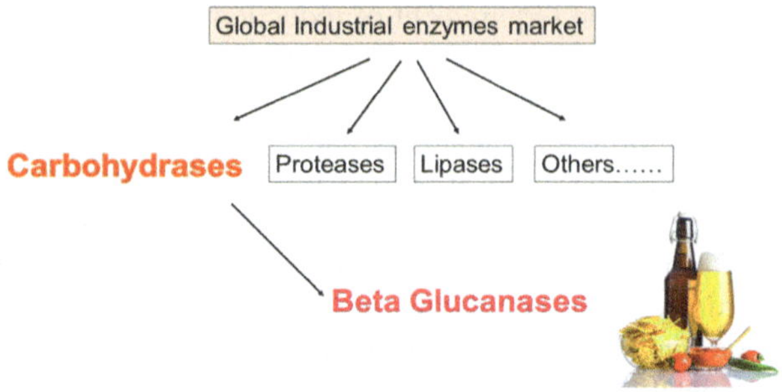

Fig. 1.1 Divergence of global industrial enzyme market

expensive food products, which is probable to increase the global demand for industrial enzymes (Dewan 2018).

The global Industrial enzymes market has diverged into carbohydrates, proteases, lipases and others (Fig. 1.1). Due to the increasing global demand for food and beverages, carbohydrases have estimated the largest share in industrial enzyme markets. Carbohydrases are glycosyl hydrolases, used in the hydrolysis of carbohydrates, to simple sugars, makes useful extensive application in industrial processes such as food, beverages, etc. (Contesini et al. 2013). Carbohydrases viz. amylase, cellulase, lactase and pectinase are commonly used in the baking industry, fruit juice and winemaking, cheese manufacturing, etc. Also have other applications viz. biofuel production, textile, detergent, leather, animal feed and pharmaceutical industries (Kirk et al. 2002). According to Technavio's market research analysis, carbohydrases have been predicted to achieve considerable market growth during 2018–2022 and are estimated to attain a CAGR close to 7% by 2022 (Technavio 2018). According to the market data forecast report, in 2021, the global carbohydrases market was evaluated by USD 3.2 billion in 2021 and expected to extend USD 5.4 billion in 2026. The industrial application of carbohydrases faces many challenges regarding its response to the pH and temperature alterations. Constant efforts in innovations and upgrades are expected to result in healthy and stable increase in carbohydrases output, which will aid in overcoming the obstacles now hindering worldwide industrial production.

1.2 Beta-Glucanases

Beta-glucanases (E.C. 3.2.1.x) are a group of carbohydrate enzymes and hydrolyse glycosidic bonds within the insoluble beta-glucan substrates, which are unable to digest by animals easily. The enzyme is distributed widely in nature and is mainly characterized extensively in microorganisms like bacteria, fungi and also in plants (Müller et al. 1998). The action of the enzyme on beta-glucan molecules releases D-glucose, as a carbon source. Beta-glucanase can perform as exo- and endo-hydrolases. Exo-beta-glucanase serially cut the glycosidic bonds from non-reducing ends of the beta-glucan chain, while endo-β-glucanase arbitrarily cuts internal bonds. In consequence of the basis of glycosidic bond cleavage, they are additionally categorized as beta-1,4-glucanases, beta-1,3-glucanases, beta-1,6-glucanases, beta-1,4-1,3-glucanases, etc. (Dake et al. 2004). Wide varieties of microbial beta-glucanases have exclusively been active on beta-glucan substrates and cleave the mixed linkage beta-glucan chains at different points. This is significant for picking precise substrates for the assay of each type of enzyme within this enzyme group. The well-recognized specificity and mechanism of action within the enormous substrates allow us to understand that most of them are varied from cellulases, excluding beta-1,4-glucanases.

Beta-glucans are non-starch, indigestible, partially water-soluble, linear polysaccharide molecules made of glucose units. They naturally occur in organic sources like cereal grains, bacteria, yeasts and algae. They are fibrous components and existed as (1 → 4)-(1 → 3) linked beta-D-glucopyranose units in cereals and (1 → 3)-(1 → 6) linked units in fungus (Havrlentova and Kraic 2006; Havrlentová et al. 2011). In cereals, beta-glucan is a principal constituent of the sub-aleurone layer and endosperm cell wall. Main sources of cereal beta-glucans are barley and oats (Wood et al. 1983). Much lower levels they also present in cell walls of wheat, rye, maize, rice, sorghum and millet (Fincher and Stone 1986; Lee et al. 1997).

Nowadays beta-glucan rich cereal food plays numerous beneficial effects in improving human health (Newman and Newman 1991). Hull-less waxy barley contains a significantly high level of beta-glucan (Ehrenbergerová et al. 2008) and is productively used in food, feed and beverages industries. It exerts an immune-modulatory positive effect in animals. In humans, it reduces cholesterol levels, coronary diseases and blood pressure (Kapur et al. 2008; Cui and Wang 2009). It can act as a substitution of dietary antibiotics in animal feeds (Moon et al. 2016). Beta-glucan rich barley is the prime constituent in breweries, which enhances colour, flavour and nutritional aspects in beer (Gupta et al. 2010). Apart from the positive effects of beta-glucan polysaccharides, studies have also been conducted relating to the negative properties. It weakens the digestion and energy metabolism in animals such as cattle and poultry. Also generates numerous problems in the brewing industry including ineffective filtration, gels and hazes formations in beer which in turn affect the taste and flavour of beer (Jin et al. 2004). The marked effect of beta-glucanase enzymes and their applications in particular industrial sectors can overcome the negative effects of beta-glucan molecules. The currently existing enzymes are not capable of the ample depolymerization of beta-glucan. In cereal beta-glucan, around 70% of glycosidic linkage is beta-(1 → 4) and the remaining 30% is beta-(1 → 3). Therefore, the complete chemical breakdown of cereal beta-glucan polymer requires the interactive action of different beta-glucanases (Fig. 1.2).

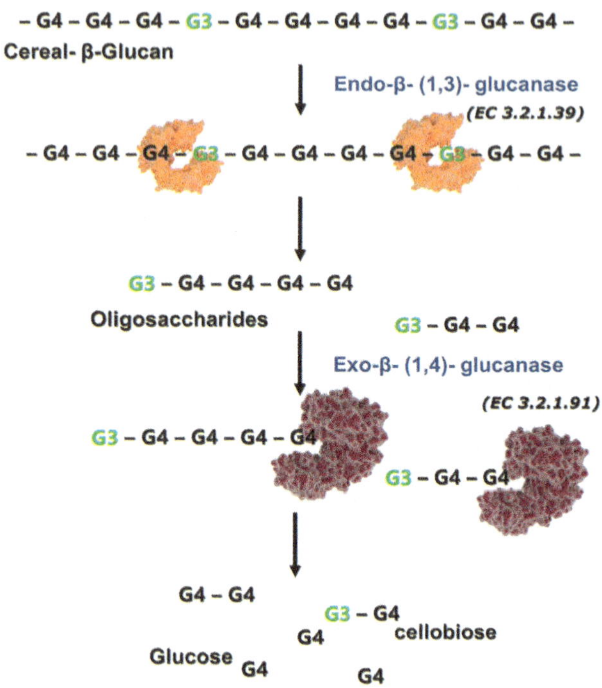

Fig. 1.2 Complete hydrolysis of cereal beta-glucan by the synergic actions of exo-beta-1,4-glucanase and endo-beta-1,3-glucanase

1.3 Biotechnological Applications of Beta-Glucanase

Beta-glucanases have noticeable ecological implications and commercial practices in well-established biotechnological industrial sectors (Fig. 1.3) like brewing and wine production, animal feed industry, textile industry, coffee processing, biofuel production, waste treatments and also in agriculture (Edison et al. 2018; Annamalai et al. 2016). In the brewing industry, it competently reduces the β-glucan stuffing in cereal grains and thereby intensifies the wort filtration rate, diminishes the chance of β-glucan precipitation and also declines the mash thickness and turbidity (Celestino et al. 2006). During winemaking, the adding of beta-glucanase progresses colour extraction, must clarification, skin softening, filtration, and finally the wine quality or superiority and stability (Singh et al. 2007). In animals and poultries, the beta-glucan rich cereal feeds cause a viscose digesta, which obstructs nutrient release and nutrient absorption in animals. Thus, the incorporation of beta-glucanases in barley rich animal feed diminishes the anti-nutritive impact on beta-glucan, in that way reducing the digesta viscosity, increasing digestibility, feed passage rates and nutrient release (Mathlouthi et al. 2002; Choct 2001). Beta-glucanases are used in the textile industry for softening the cotton clothing, for giving a finishing look to the denim materials, for the colour care and cleaning of clothing (Cherry and Fidantsef 2003). Beta-glucanase is applied as a biocontrol agent in agriculture sectors because it is proficient to degrade the cell wall of plant pathogenic fungi. Beta-glucanase

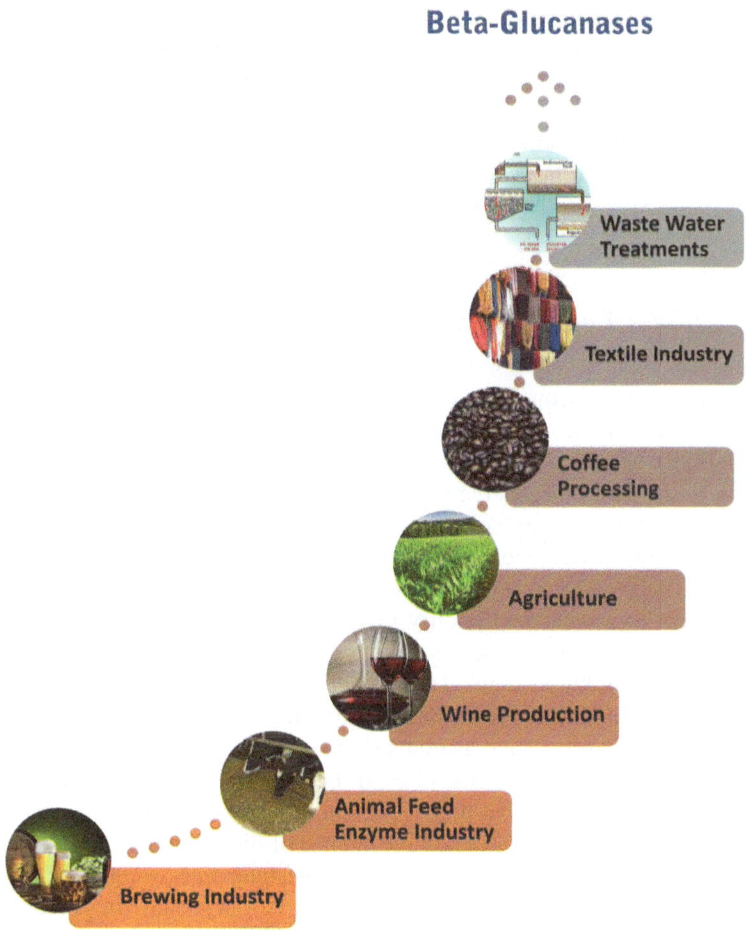

Fig. 1.3 Applications of beta-glucanases in various industrial sectors

enhances seed germination by weakening the seed endosperm. This enzyme also helps for the preparation of plant and fungal protoplasts relevant in various research purposes (Bhat 2000).

1.3.1 Beta-Glucanase in Brewing Industry

Beta-glucanases are broadly used in the brewing industry, there it permits the beta-glucan degradation of cereal grain such as barley and wheat. Barley contains 4–7% of beta-glucan, present in cell wall endosperm. The soluble or partially soluble beta-glucan content causes viscosity. During the brewing process, the excess beta-glucan content in barley and less modified mash creates high wort viscosity, thereby creating filtration problems. Another problem related to beta-glucans in the brewery is gel and haze formation in cold beer, this also causes difficulties in final beer filtration. The addition of beta-glucanase added at the time of mashing helps to improve the wort filterability and also reduces haze formation, thereby it improves the yield and quality of beer and gives economically beneficial brewery production (Briggs et al. 2004).

1.3.2 Animal Feed Enzyme Industry

Animal feeds with viscous cereals like wheat, barley, rye or triticale contain a relatively large proportion of soluble and insoluble fibres and non-starch polysaccharides (NSP). Monogastric animals such as pigs and poultry are unable to digest fibres due to the lack of digestive enzyme production. The most prominent soluble fibre is beta-glucan. A major source of beta-glucan in animal feeds is barley grain. Oats also contain certain levels of beta-glucan and these are infrequently served to chickens and pigs. Water-soluble, viscous beta-glucans have a lethal effect in livestock varieties particularly poultry (Walsh et al. 1995). They increase the viscosity of the fillings in the small intestine to significantly disturb the rate of movement of barley-based diets. This hinders the digestion of nutrients and their absorption, consequently reducing animal growth. The problem is also linked with certain digestive disorders such as non-specific colitis in swine and hock burns and sticky litter in poultry. The addition of beta-glucan degrading enzyme beta-glucanase can effectively solve these problems by improving the digestive performance of the feeds and also decrease the chance of digestive disorders (Mathlouthi et al. 2003). Beta-glucanase action reduces intestinal tract weights, changes the intestinal morphology and microflora profiles in the large and small intestine and provides overall animal health. As the enzyme utilizes the substrate present in the gut, there is a direct impact on the microfloral populations. This helps uplifting the growth of poorly performing animals and keeps uniformity in the market.

1.3.3 Wine Production

In wine production, polysaccharides such as beta-glucans, chitin, cellulose, etc. originate from the autolysis of yeast cell walls, the grape berries, some species of lactic acid bacteria particularly *Pediococcus* spp. and the grape fungus *Botrytis cinerea*. These sticky, colloidal extracellular polysaccharides harm the wine filtration because beta-glucans are very difficult to remove through adsorbents, flocculants or by filtration (Claus and Mojsov 2018). The application of different beta-glucanase during enological processing is useful for reducing viscosity in wine, thus improving the mouthfeel flavour and colour of wines.

1.3.4 Biocontrol Activity

Cell walls of pathogenic fungi like *Sclerotium rolfsii*, *Rhizoctonia solani*, etc. are predominantly constituted by polysaccharides β-glucan, which comes after chitin. Some evidence proposes that beta-glucan plays a prominent role in fungal pathogenesis in plants (Ruel and Joseleau 1991). The enzyme glucanase also known as pathogenesis-related protein (PR), kills pathogenic fungi and results in the loss of cell contents by cleaving the beta-glucan polysaccharides of the fungus. Synergistic action of different beta-glucanases needs complete degradation of the beta-glucans cell wall. Beta-glucanases from microbial origin have been widely reported to accomplish extensive crop protection and also promote plant growth (Gopalakrishnan et al. 2013; Mouyna et al. 2013). Streptomyces species are also frequently used for producing lysing enzymes against a large spectrum of pathogenic fungi. An important factor of actinomycetes fungus antagonism is the extracellular enzymes having beta-glucanase activities. Streptomyces sp. EF-14 produced beta-1,6-glucanase and beta-1,3-glucanase have been recognized as the best effective fungal antagonist of Phytophthora spp. Soil supplemented with *Streptomyces nigellus* NRC 10 strain reduced the percentage of damping-off diseases in tomato plants caused by *Pythium ultimum*. Studies indicated that the strains are excellent producers of the β-1,3 and β-1,4-glucanases (Helmy et al. 2010).

1.3.5 β-Glucanase in Protoplast Preparation

Protoplasts, the naked plant cells, are the most commonly used research tool in plant cells for parasexual modification and provide numerous benefits for crop improvement. Due to the high specificity and discriminating hydrolysis of cell wall polymers, beta-glucanases are widely applied in the protoplast preparations from plant, bacterial and fungal cells. Protoplasts are valuable in numerous areas of research such as for isolating high molecular weight DNA from cells with the tough cell wall, transformation, propagating wild plant species using tissue culture, etc. For successful protoplast preparation, an adequate cell wall must be removed to allow fusion between protoplasts and a sufficient cell wall must remain for the successful regeneration of the complete cell wall (Ezeronye and Okerentugba 2001). Nowadays different beta-glucanase formulations are commercially available. Some are scientifically proved as best for the preparation of protoplasts. Zymolyase® and Novozym 234 is a commercially available enzyme preparation of a mixture of beta-glucanase which has been used to prepare protoplasts (Jung et al. 2000). Yatalase, prepared from Corynebacterium culture supernatants, is used for preparing protoplasts from filamentous fungi. It consists mainly of combinations of chitinase, chitobiose and beta-1,3-glucanase.

1.3.6 Germination Enhancement by Beta-Glucanases

Recently explored the new function of beta-glucanases was endosperm weakening as part of seed germination. Endosperm weakening or dissolution of the mechanical resistance of the micropylar endosperm seems to be necessary for the successful completion of seed sprouting. According to Ikuma and Thimann (1963), the production of enzymes is the ultimate phase in regulating the germination process, which helps to facilitate the penetration of the tip of the radicle through the coat. Leubner-Metzger (2003) proposed that class beta-1,3-glucanase is induced within the micropylar endosperm before its rupture and is closely accompanying by altered endosperm rupture in the regulation of germination in response to light, gibberellins, abscisic acid (ABA) and ethylene. These findings support that cell-wall-modifying proteins such as enzyme beta-1,3-glucanase have significant contributions in the regulation of various plant physiological and developmental processes like microsporogenesis, pollen germination, pollen tube growth, fertilization, embryogenesis, seed development, seed germination, mobilization of endosperm, cereal grain storage reserves, dormancy release and after-ripening of dicot seeds.

Microbial beta-glucanases are considered environmentally safe resolutions for many environmentally dangerous complications. They may also be used as a sophisticated tool in beta-glucan structural analysis. The most notable problem associated with this study is the commercial accessibility substrates of beta-glucanase and the market cost. Only some pure beta-glucans products are available from some companies that can be utilized as the substrate for beta-glucanases and very expensive, which in turn increases the overall production outlays of the beta-glucanase enzymes. The increased cost of enzyme limits its applications in large-scale industries. As a part of the extensive production, some of the beta-glucanase genes have been effectively cloned and expressed in heterologous host cells, this also facilitated a better understanding of the regulation of enzyme synthesis. In addition, the genetic engineering approach is a powerful tool to construct recombinant expression cassettes encoding different beta-glucanase genes and the production of enzymes on a commercially reasonable scale. At present only a few organisms are exploited for screening beta-glucanase enzyme systems. Further research for the purification and characterization of more enzymes will be necessary for a better understanding of their properties. The use of cost-effective microbial enzymes is an efficient tool for maintaining a green environment. Even though the beta-glucanase has many applications in various sectors, currently major portion is utilized in brewing as well as animal feed enzyme industries for hydrolysing barley beta-glucan. Beta-glucanase also has been identified as one of the most demanding enzymes in the global enzyme market when considering the fastest growth and future needs in beverage and animal feed enzyme industries.

1.4 Beta-Glucanases: Market Dynamics

The ever-growing beverages market by the popularity of speciality beers and craft breweries, the β-glucanase segment is predicted to be a significant rise in demand and as the fastest emergent for brewing enzymes from 2018 to 2023. The market of brewing enzymes is projected at $352.1 million in 2018 and is projected to reach $484.7 million by 2023, at a CAGR of 6.6% from 2018 (Globe Newswire 2018). The fastest-growing brewing enzymes market is projected to be in the Asia Pacific region and is expected to witness significant demand in future also. Maximum growth is expected to be seen exactly from India and China. Substantial growth was also observed in Africa and Latin America (Future market insights 2019). The global feed enzyme market is a fast emergent segment due to increasing consumption, rising the quality and safety of meat and milk and attention to favourable feeding plans for livestock. The market was valued at USD 899.19 million in 2014 and is estimated to stretch USD 1371.03 million by 2020, at a CAGR of 7.3% and projected to reach USD 1.9 billion by 2025 with a CAGR of 8.1% (Wattagnet 2015). The principal concern of the market is animal health and also the high nutrient uptake of feed. The steady growth is predominantly higher in developing countries such as India, China and Brazil owing to the increase in the returns level and attention towards superior quality, nutritive and balanced food grades. In terms of value, the major share in the feed enzyme market is contributed by non-starch polysaccharide enzymes (NSP), in which the β-glucanase is a leading segment (Research and Markets 2013). The demand for healthier and highly efficient feed enzymes with modified capabilities is an alternative to feed antibiotics and also manage the input production cost of livestock.

The microbial source of beta-glucanase enzymes dominated as major share in the brewing and feed enzymes market as they have a high growth rate, easy to handle and can be manufactured in huge tanks without illumination. Uninterrupted R&D as well as resulting advancements in processes and product innovations in these areas rising global per capita incomes. Future industrial application of beta-glucanase is expected to explore the natural resources of enzymes and the development of new beta-glucanase for improving food products by genetic engineering.

References

Annamalai N, Rajeswari MV, Balasubramanian T (2016) Endocrinologist-1,4. β-glucanases: role, applications and recent developments. In: Gupta V (ed) Microbial enzymes in bioconversions of biomass. Biofuel and biorefinery technologies, vol 3. Springer, Cham, pp 37–45

BCC research (2021) Global markets for enzymes in industrial applications. Available from https://www.bccresearch.com/market-research/biotechnology/global-markets-for-enzymes-in-industrial-applications.html

Bhat MK (2000) Cellulases and related enzymes in biotechnology. Biotechnol Adv 18(5):355–383

Briggs DE, Boulton CA, Brookes PA, Stevens R (2004) Brewing: science and practice: Cambridge UK: Woodhead publishing/Boca Raton. CRC Press, FL, USA, p 881

Celestino KRS, Cunha RB, Felix CR (2006) Characterization of a β-glucanase produced by Rhizopus microsporus var. microsporus, and its potential for application in the brewing industry. BMC Biochem 7:7:23

Chapman J, Ismail AE, Dinu CZ (2018) Industrial applications of enzymes: recent advances, techniques, and outlooks. Catalysts 8(6):238

Cherry JR, Fidantsef AL (2003) Directed evolution of industrial enzymes: an update. Curr Opin Biotechnol 14(4):438–443

Choct M (2001) Enzyme supplementation of poultry diets based on viscous cereals. In: Bedford M, Partridge G (eds) Enzymes in farm animal nutrition. CABI Publishing, Oxon, pp 145–160

Choi JM, Han SS, Kim HS (2015) Industrial applications of enzyme biocatalysis: current status and future aspects. Biotechnol Adv 33(7):1443–1454

Claus H, Mojsov K (2018) Enzymes for wine fermentation: current and perspective applications. Fermentation 4(3):52

Contesini FJ, de Alencar FJ, Kawaguti HY, de Barros Fernandes PC, de Oliveira CP, da Graça NM, Sato HH (2013) Potential applications of carbohydrases immobilization in the food industry. Int J Mol Sci 14(1):1335–1369

Cui SW, Wang Q (2009) Cell wall polysaccharides in cereals: chemical structure and functional properties. Struct Chem 20(2):291–297

Dake MS, Jhadav JP, Patil NB (2004) Induction and properties of 1,3-β-D-Glucanases from Aureobasidium pullulans. Indian J Biotechnol 3:58–64

Dewan SS (2018) Global markets for enzymes in industrial applications; BCC Research: Wellesly: MA, USA. Available from https://www.persistencemarketresearch.com/market-research/industrial-enzymes-market.asp

Edison LK, Shiburaj S, Pradeep NS (2018) In: Kumar P et al (eds) Microbial Beta Glucanase in agriculture. Microbial advances in agriculture and human health. Apple academic press, Inc., (a Taylor & Francis Group), Canada, pp 53–86. https://doi.org/10.1201/9781351248914-10

Ehrenbergerová J, Brezinová Belcredi N, Psota V, Hrstková P, Cerkal R, Newman CW (2008) Changes caused by genotype and environmental conditions in beta-glucan content of spring barley for dietetically beneficial human nutrition. Plant Foods Hum Nutr 63(3):111–117

Ezeronye OU, Okerentugba PO (2001) Optimum conditions for yeast protoplast release and regeneration in Saccharomyces cerevisiae and Candida tropicalis using gut enzymes of the giant African snail Achatina achatina. Lett Appl Microbiol 32(3):190–193

Fincher GB, Stone BA (1986) Cell walls and their components in cereal grain technology. In: Pomaranz Y (ed) Advances in cereal technology, vol 8. AACC, St. Paul, pp 278–295

Future market insights (2019) Beta-glucanase market: global industry analysis and opportunity assessment 2016–2026. Retrieved from https://www.futuremarketinsights.com/reports/Beta-glucanase-market

Globe Newswire (2018) Global brewing enzymes market 2018–2023: the beta-glucanase segment is projected to register the fastest growth. Retrieved from https://globenewswire.com/newsrelease/2018/08/07/1547839/0/en/Global-Brewing-Enzymes-Market-2018-2023-The-Beta-Glucanase-Segment-is projected-to-Register-the-Fastest-Growth.html

Gopalakrishnan S, Srinivas V, Sree Vidya M, Rathore A (2013) Plant growth-promoting activities of Streptomyces spp. in sorghum and rice. SpringerPlus 2(1):574

Gupta M, Abu-Ghannam N, Gallaghar E (2010) Barley for brewing: characteristic changes during malting, brewing and applications of its by-products. Compr Rev Food Sci Food Saf 9(3): 318–328

Gurung N, Ray S, Bose S, Rai V (2013) A broader view: microbial enzymes and their relevance in industries, medicine and beyond. Biomed Res Int 329121:1–18

Havrlentova M, Kraic J (2006) Content of β-D-glucan in cereal grains. J Food Nutr Res 45(3): 97–103

Havrlentová M, Petruláková Z, Burgárová A, Gago F, Hlinková A, Šturdík E (2011) Cereal β-glucans and their significance for the preparation of functional food–a review. Czech J Food Sci 29(1):1–14

Helmy SM, Zeinat K, Mahmoud SA, Moataza S, Nagwa M, Amany H (2010) Streptomyces nigellus as a biocontrol agent of tomato damping-off disease caused by Pythium ultimum. Electron J Pol Agric Universities S Biotechnol 13:4

Ikuma H, Thimann KV (1963) The role of the seed coats in germination of photosensitive lettuce seeds. Plant Cell Physiol 4:169–185

Illanes A (2008) Enzyme biocatalysis–principles and applications. Springer, New York, NY

Jin YL, Speers RA, Paulson AT, Stewart RJ (2004) Barley beta-glucans and their degradation during malting and brewing. Tech Q Master Brew Assoc Am 41:231–240

Jung MK, Ovechkina Y, Prigozhina N, Oakley CE, Oakley BR (2000) The use of Beta-D-glucanase as a substitute for Novozym 234 in immunofluorescence and protoplasting. Available from http://www.fgsc.net/fgn47/47Jung.htm

Kapur NK, Ashen D, Blumenthal RS (2008) High density lipoprotein cholesterol: an evolving target of therapy in the management of cardiovascular disease. Vasc Health Risk Manag 4(1): 39–57

Kirk O, Borchert TV, Fuglsang CC (2002) Industrial enzyme applications. Curr Opin Biotechnol 13(4):345–351

Lee CJ, Horsley RD, Manthey FA, Schwarz PB (1997) Comparisons of β-glucan content of barley and oat. Cereal Chem 74(5):571–575

Leubner-Metzger G (2003) Functions and regulation of β-1,3-Glucanases during seed germination, dormancy release and after-ripening. Seed Sci Res 13(1):17–34

Liese A, Seelbach K, Wandrey C (2006) Industrial biotransformations, 2nd. Liese et al., eds.: Weinheim: Wiley-VCH, pp. 273–447

Liu Z, Tyo KE, Martínez JL, Petranovic D, Nielsen J (2012) Different expression systems for production of recombinant proteins in Saccharomyces cerevisiae. Biotechnol Bioeng 109(5): 1259–1268

Market data forecast report (2021) Carbohydrase Market. Available from https://www.marketdataforecast.com/market-reports/global-carbohydrase-market

Mathlouthi N, Juin H, Larbier M (2003) Effect of xylanase and β-glucanase supplementation of wheat- or wheat- and barley-based diets on the performance of male turkeys. Br Poult Sci 44(2): 291–298

Mathlouthi N, Mallet S, Saulnier L, Quemener B, Larbier M (2002) Effects of xylanase and beta-glucanase addition on performance, nutrient digestibility and physico-chemical conditions in the small intestine contents and caecal microflora of broiler chickens fed a wheat and barley-based diet. Anim Res 51(5):395–406

Moon SH, Lee I, Feng X, Lee HY, Kim J, Ahn DU (2016) Effect of dietary beta-glucan on the performance of broilers and the quality of broiler breast meat. Asian-Australas J Anim Sci 29(3): 384–389

Mouyna I, Hartl L, Latgé JP (2013) β-1,3-glucan modifying enzymes in aspergillus fumigatus. Front Microbiol 4(81):81

Müller JJ, Thomsen KK, Heinemann U (1998) Crystal structure of barley 1,3-1,4-beta-glucanase at 2.0-a resolution and comparison with bacillus 1,3-1,4-beta-glucanase. J Biol Chem 273(6): 3438–3446

Newman RK, Newman CW (1991) Barley as a food grain. Cereal Foods World 36:800–805

Pandey A, Binod P, Ushasree MV, Vidya J (2010) Advanced strategies for improving industrial enzymes. Chemical Industry Digest 23:74–84

Poulsen PB, Bucholz K (2003) History of enzymology with emphasis on food production. In: Whitaker JR, Voragen AGJ, Wong DWS (eds) Handbook of food enzymology: New York. Marcel Dekker, NY

Research and Markets (2013) Feed enzyme market by type (phytase, protease & nsp), sub-type (xylanase, cellulase, pectinase, â-glucanase & mannase), & application (swine, poultry, ruminant, aqua feed & others)-global trends & forecasts to 2018. Retrieved from https://www.researchandmarkets.com/reports/2556103/feed_enzyme_market

Ruel K, Joseleau JP (1991) Involvement of an extracellular glucan sheath during degradation of populous wood by Phanerochaete chrysosporium. Appl Environ Microbiol 57(2):374–384
Singh A, Kuhad RC, Ward OP (2007) Industrial application of microbial cellulases. In: Kuhad RC, Singh A (eds) Lignocellulose biotechnology: future prospects. I. K. International Publishing House, New Delhi, India, pp 345–358
Singh R, Kumar M, Mittal A, Mehta PK (2016) Microbial enzymes: industrial progress in 21st century. 3. Biotech 6(2):174
Technavio (2018) Global carbohydrases market 2018–2022. Available from https://www.technavio.com/report/global-carbohydrases-market-analysis-share-2018
Walsh GA, Murphy RA, Killeen GF, Headon DR, Power RF (1995) Technical note: detection and quantification of supplemental fungal beta-glucanase activity in animal feed. J Anim Sci 73(4): 1074–1076
Wattagnet (2015) Global feed enzymes market to reach $1,371M by 2020. Retrieved from https://www.wattagnet.com/articles/22536-global-feed-enzymes-market-to-reach-1-371m-by-2020
Wood PJ, Fulcher RG, Stone BA (1983) Studies on the specificity of interaction of cereals cell wall components with Congo red and Calcofluor: Specific detection and histochemistry of (1→3), (1→4)- β-D-glucan. J Cereal Sci 1(2):95–110

Chapter 2
Structure and Classification of Beta-Glucanases

Lekshmi K. Edison, Karthik Menon, and N. S. Pradeep

Abstract Beta-glucanases are clustered under the family glycosyl hydrolase (GH). The assorted roles of beta-glucanases in nature are completely determined by the various chemical interactions and molecular structures. Therefore, sophisticated systems are essential for their industrial level production and processing. The clarification and understanding of the molecular structure of beta-glucanases benefit to recognize the molecular mechanism and cleavage pattern of a wide array of beta-glucan substrates. Modes of action of enzymes were scientifically scrutinized based on the sequence similarity, phylogeny studies, kinetics studies, mutagenesis, X-ray crystallography and molecular dynamics. Definitions of molecular structure and catalytic mechanism of beta-glucanase help to expand the glucanolytic research from basic understanding to high throughput biotechnological applications with the aid of genetic engineering and protein engineering technologies.

Keywords Glycosyl hydrolase · Catalytic mechanism · Carbohydrate-binding module

2.1 Introduction

Catalytically active enzymes are responsible for the biological processes in living cells (Agarwal 2006). The catalytic functions of enzymes fundamentally depend on the three-dimensional structure. Scrutinizing the structure of an enzyme expands the understanding of the structural role in enzyme function and catalytic performance. The structural perceptions and understanding of catalytic roles also support the

---

L. K. Edison
Rajiv Gandhi Centre for Biotechnology (RGCB), Thiruvananthapuram, Kerala, India

K. Menon
Membrane Biology Laboratory, Crystallography and Molecular Biology, Saha Institute for Nuclear Physics, Kolkata, India

N. S. Pradeep (✉)
KSCSTE-Malabar Botanical Garden and Institute for Plant Sciences, Kozhikode, Kerala, India

N. S. Pradeep, L. K. Edison (eds.), *Microbial Beta Glucanases*, Interdisciplinary Biotechnological Advances, https://doi.org/10.1007/978-981-19-6466-4_2

annotation of unknown functions, performing structural manipulations, designing new enzymes, disrupting enzyme function and predicting cellular metabolism. However, the enzymes are large molecules, the functional catalytic regions are restricted in a small part called cleft or pocket, in which the enzyme interacts with the substrate and reaction performs (Laskowski et al. 1996). Chemical groups in the active sites can be determined by NMR experiments and/or also by site-directed mutagenesis. Overall, the three-dimensional structure of the enzyme defines every functional aspect of the enzyme such as chemical catalysis, substrate binding, stability regulation, etc.

Beta-glucans are cell wall polysaccharide components rich in the endosperm cell wall of the Gramineae family, especially in barley, sorghum, rye, wheat and rice. They are linear glucans connected through beta-1,3 and beta-1,4 glycosidic linkages. The depolymerization of plant beta-glucans is done by endogenous beta-glucanase or glucanohydrolases during the germination process. Endo glucanase is glycosyl hydrolases. Plant beta-glucan depolymerization requires at least three glucanohydrolases with endo or exo mode of action: beta-1,4-glucanase, beta-1,3-glucanase and beta-1,4-1,3-glucanase. Microorganisms also secrete diverse beta-glucan degrading enzymes and use them as their energy source. All those enzymes display different and stringent substrate specificities which depend on their molecular structure. Plants and microbial beta-glucanase do not share any similarities in amino acid sequence and the associated three-dimensional structure, whereas they show the same substrate specificity as a result of convergent evolution. For example, plant 1,3-1,4-endo beta-glucanase has an 8-barrel three-dimensional structure, but the same enzyme from the bacterial source showed a jellyroll sandwich structure. Bacterial beta-glucanases have significant biotechnological benefits in brewing as well as the animal feed industry. During high temperature treatments in the malting stage of a brewery, the endogenous beta-glucanases of cereal grains become heat-inactivated. Hence, the undigested beta-glucans that remain in the malt cause severe issues such as reduction in filtration rate and yield. The gelatinous precipitates of beta-glucan affect the beer quality also. The addition of thermostable beta-glucanase from a microbial source is important for improving the filtration rate, thus enhances overall beer quality. The high beta-glucan rich animal feed also causes indigestion problems in broiler chicken and piglets. The enzymatic preparations including microbial beta-glucanase enhance the digestibility of cereal diets in animals (Stone and Clarke 1992; White et al. 1983; Planas 2000). Microbial beta-glucanases are grouped into different families based on the alignments in amino acid sequences and hydrophobic cluster analyses. The understanding of three-dimensional structure helps for unravelling the mechanism of action and enzymatic properties of beta-glucanase.

2.2 Beta-Glucanases Are Glycosyl Hydrolases (GHs)

Glycosyl hydrolases (E.C.3.2.1.x) are popular and industrially useful groups of enzymes involved in the hydrolysis of glycosidic bonds within carbohydrate molecules or between carbohydrate and non-carbohydrate molecules. They are well-studied groups of enzymes with wide range of applications in the food industry, biomass degradation, bioethanol production, waste processing and pest control (Fulop and Ponyi 2015). GHs are extensively distributed prokaryotic, eukaryotic, and also in archaea with multiple copy numbers in each organism. GHs are categorized primarily based on the substrate specificity and further classified according to the sequence similarities, catalytic mechanisms and hydrophobic cluster analysis.

The first GH structure solved more than 30 years ago, was hen egg-white lysozyme. Currently, the GHs are grouped into 140 families (Cantarel et al. 2009; Abe et al. 2017), deliver an understanding of the comparative structural features, evolutionary relationships and mechanisms of action (Sathya and Khan 2014). The updated list of GH families is available in the Carbohydrate-Active Enzymes database (CAZy). GHs have a segmental structure (Fig. 2.1) comprising a catalytic domain and one or two non-catalytic domains, some of which, like the carbohydrate binding domain, are involved in the substrate binding (Davies and Henrissat 1995). The active site topologies of GHs are categorized into three classes—pocket or crater, cleft or groove and tunnel (Fig. 2.2).

Pocket or *crater*—It recognizes the non-reducing extremity of a saccharide, hydrolyses mainly the amorphous polymers such as starch and dimers like cellobiose and they are not very efficient for the hydrolysis of cellulose like fibrous substrates.

Cleft or *groove*—An 'open' structure randomly binds the sugar units in polysaccharides, usually found in endo-acting GHs such as endocellulases, lysozymes, chitinases, xylanases, α-amylases, β-1,3-glucanases and β-1,3-1,4-glucanases.

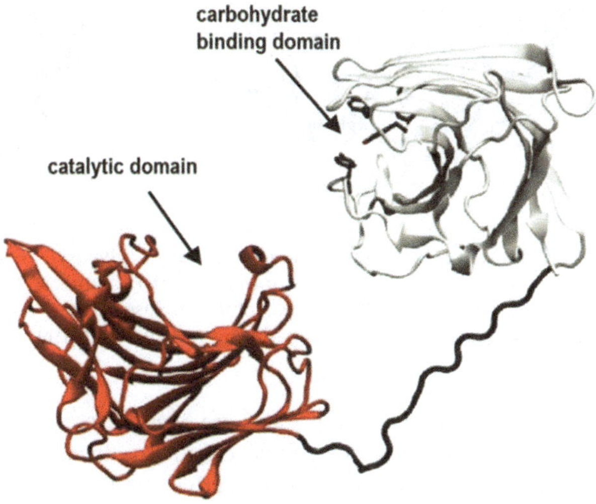

Fig. 2.1 A representative structure of GH family protein (Codera et al. 2015)

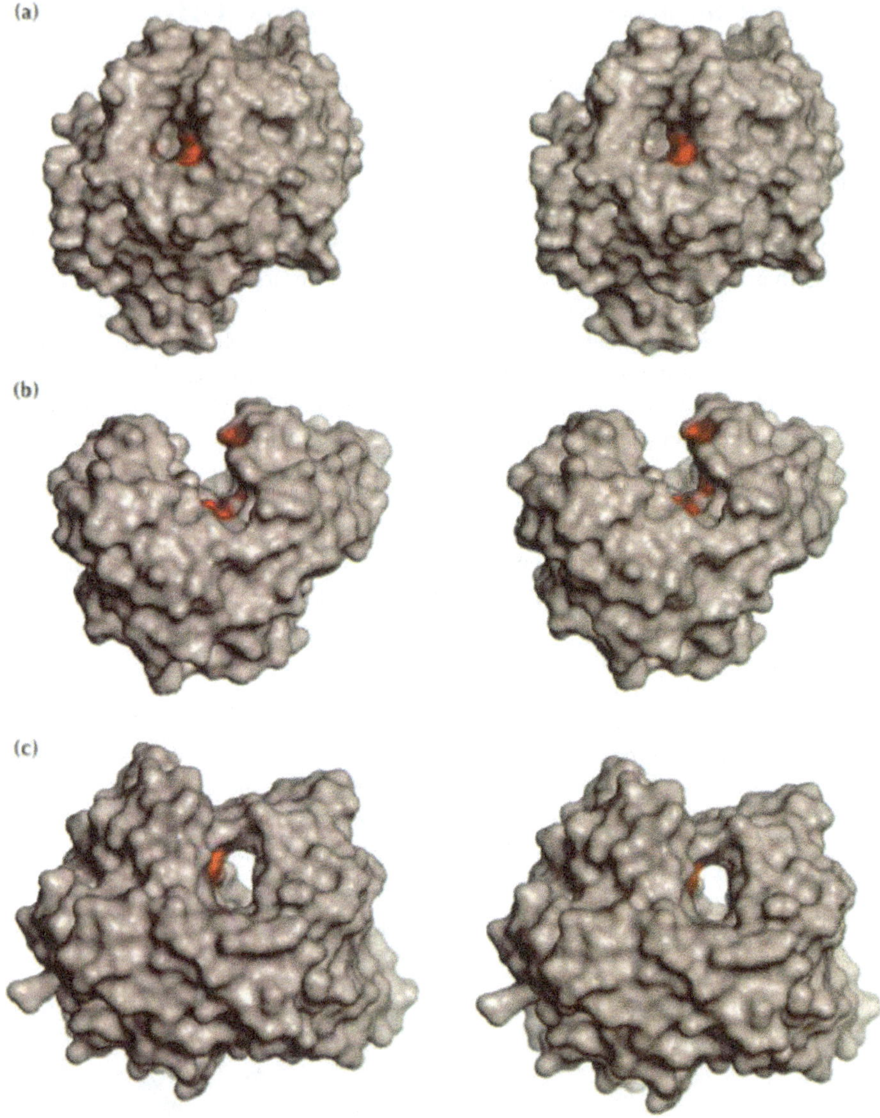

Fig. 2.2 Active site configurations found in glycosyl hydrolases. (**a**) Pocket or crater (glucoamylase from *Aspergillus awamori*). (**b**) Cleft or groove (endoglucanase E2 from *Thermobifida fusca*). (**c**) Tunnel (cellobiohydrolase II from *Trichoderma reesei*). The predicted catalytic residues are shown in red (Davies and Henrissat 1995)

Tunnel—The catalytic pockets are encompassed within the tunnel and allow the progressive hydrolysis of cellulose and exoglucanase like polymers. It allows the polysaccharide chain to enter the active site from one entrance of the tunnel in an 'exo' fashion and release the products from the other side while firmly clamping the chain.

2.3 Catalytic Mechanism

Regardless of the enormous number of GH families, they have a common catalytic mechanism, either an inversion or retention of an anomeric configuration. The hydrolysis occurs through general acid/base catalysis, by the action of two essential residues: a proton donor and a nucleophile or base residue. The first one is commonly glutamate or aspartate, while the second is glutamic or aspartic acid (an exception is sialidases, where activated tyrosine is nucleophile). Inverting GHs work through a single nucleophilic substitution, whereas the retaining enzymes follow a double-displacement mechanism through making and hydrolysing covalent intermediate (Fig. 2.3). In both mechanisms, each step contains a sp^2 hybridized oxocarbenium ion-like transition state (TS), and a positive charge is developed at anomeric carbon. The electron donated by ring oxygen partially stabilizes the positive charge (Ardèvol and Rovira 2015).

In inverting mechanism, a nucleophilic attack is performed by an activated water molecule at the C-1 position of the sugar, along with a concomitant departure of

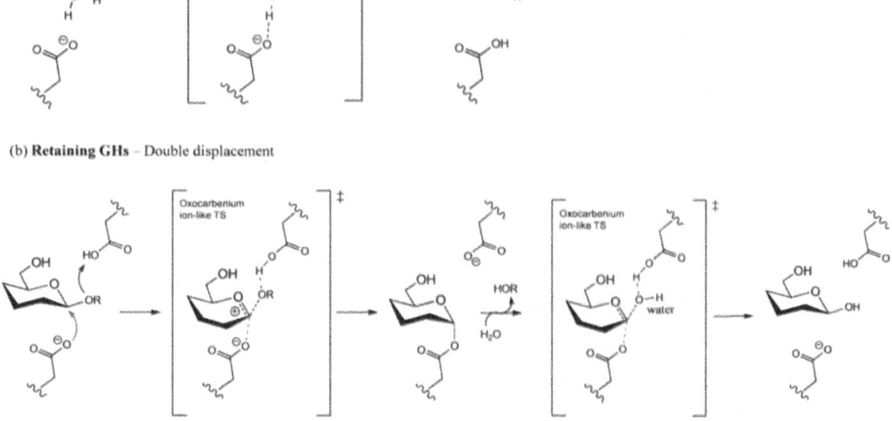

Fig. 2.3 Catalytic mechanisms of glycosyl hydrolases (GHs). (**a**) Inverting GHs and (**b**) retaining GHs (Ardèvol and Rovira 2015)

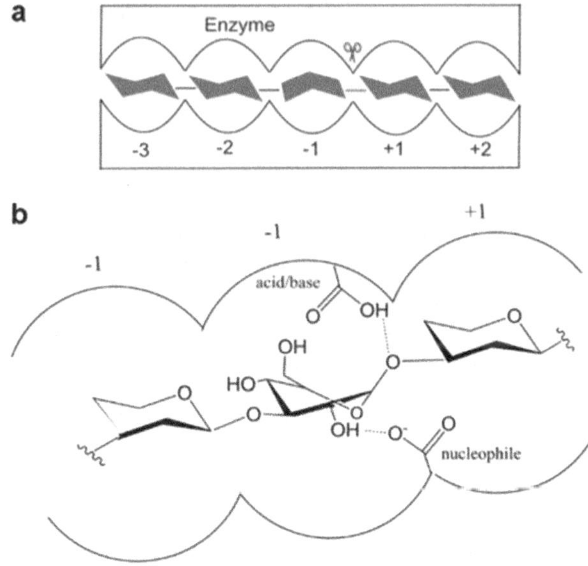

Fig. 2.4 Sugar distortion in GHs. (**a**) The saccharide binding in GH. The enzyme cleaves the bond located in between −1 and + 1 sub-sites. In the "catalytic" −1 sub-site the sugar adopts a distorted conformation like boat or skew-boat, while other sugar moieties have a relaxed chair like confirmation. (**b**) Sugar distortion at the −1 sub-site. (Ardèvol and Rovira 2015)

aglycone which is done by glycosidic oxygen protonation. In contrast, a covalent GH intermediate is involved in the retaining mechanism. In the first step, glycosidic oxygen is protonated by one carboxyl group which functions as a general acid and makes a concomitant bond cleavage. Then a covalent intermediate is formed by the nucleophile attack of the second carboxyl group. In the second step, the inward water molecule is activated by a deprotonated acid catalyst by acting as a base. The activated water molecule executes the second nucleophilic attack at the anomeric carbon of sugar, thus releasing free sugar with the same anomeric stereochemistry as the substrate (Nishimura et al. 2001). The substrates modify the conformation to a distorted form when binding to the GHs. Particularly, the saccharide unit attaching the catalytic site of the enzyme habitually changes to a skew-boat or boat type conformation (Espinosa et al. 1998) (Fig. 2.4).

According to Teegardin et al. (2017), different GHs family proteins are widely distributed in the Actinobacteria family, especially in *Streptomyces* species. Attention to understanding GHs 3-dimensional structures is highly significant in the view of the application and also towards protein engineering.

2.4 Carbohydrate-Binding Modules (CBMs)

GHs are modular enzymes, consisting of catalytic modules and carbohydrate-binding modules (CBMs). The CBMs act like a clamp that helps to focus the enzyme on the polysaccharide substrates to improve the degradation efficiency. The structure and binding specificities of several CBMs have been reported (McLean et al. 2002).

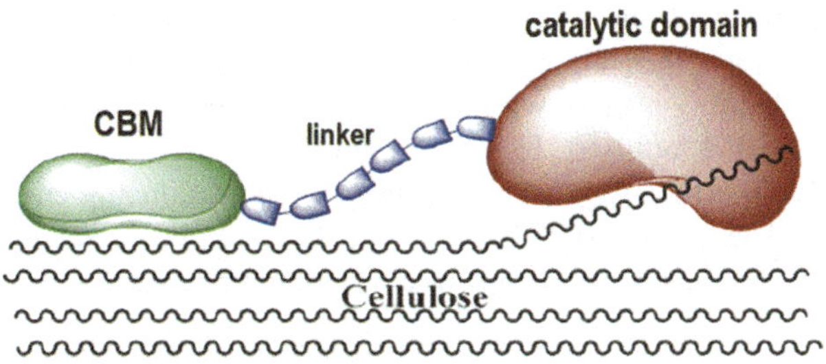

Fig. 2.5 Diagrammatic representation of the GHs action on cellulose fibres (Sajith et al. 2016).

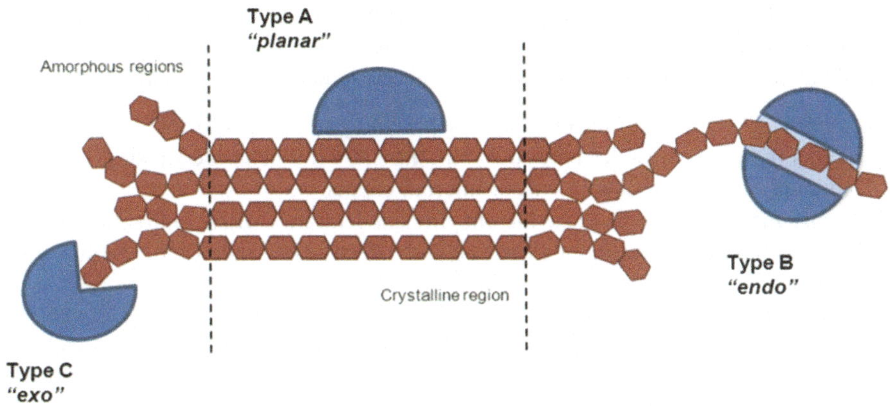

Fig. 2.6 Schematic representation of CBM types showing the affinity to diverse regions of a polysaccharide substrate (Baroroh et al. 2017)

Currently, there are 84 CBM families, which are recognized to have diverse properties. Within the family, CBMs showed similar structural folding and have quite dissimilar specificities to substrate binding (Abou Hachem et al. 2000). Additionally, CBMs provide some other functions, for example, the members from CBM family 2 bind to crystalline cellulose and help to disorder the crystalline architecture, thus increasing the accessibility of enzyme to the substrate (Din et al. 1991) (Fig. 2.5). The position of CBM in carbohydrases can be both N-, C- terminal or centrally. Based on the nature and degree of polymerization to the substrates, the CBMs are categorized into three types (Armenta et al. 2017) (Fig. 2.6).

Type A: Binds to the hydrophobic regions of highly crystalline polysaccharide surfaces. They are the members of the CBM family 1, 2, 3, 5 and 10. They are also reported to have the affinity towards soluble polysaccharides (especially CBM2 and CBM3). Binding sites have flat or planar hydrophobic architecture, rich with aromatic amino acid residues which play a key role in substrate binding.

Type B: Endo-type CBM, mainly found in the C-terminus of laminarinase (endo-β-1,3-glucanase) for recognizing the β-1,3-linked saccharide units, and also recognize substrates like β-1,3-glucans, β-1,4-mannan, mixed β-(1,3)(1,4)-glucans, galactomannan and glucomannan. The binding involves 3 tryptophan residues and 1 tyrosine residue (Boraston et al. 2002). The binding sites look like grooves or clefts and accommodate four or more sugar units. It exhibits high binding affinity with increasing the substrate chain length. This type includes the CBM family 6, 13, 20, 36 and 60.

Type C: Exo-type and recognizes the non-reducing end of polysaccharide substrate. The families of CBM9, 13, 32, 47, 66 and 67 come under this type. Type C CBMs are considered 'lectin-like' (Datta and Sapra 2011).

CBMs are accessory modules that assist the catalytic modules in the degradation of complex polysaccharides. It is also involved in protein toxin delivery, bacterial host cell colonization and glycan synthesis. The occurrence of CBM in GHs is a key factor for increasing the ability of efficient hydrolysis of insoluble polysaccharides. CBMs have four main functional roles during polysaccharide degradation:

Targeting Effect: Depending on the enzyme catalytic site conformation the CBM targets the enzyme to the particular binding site on saccharide substrate such as reducing end, non-reducing end or internal chains.

Proximity Effect: CBMs increase the enzyme concentration close to the binding site of a saccharide substrate, increase fast and effective degradation of the substrate (Herve et al. 2010).

Disruptive Effect: CBMs help to disrupt the tightly packed surface of polysaccharides, which loosen the substrates and expose more to the catalytic module of enzymes for effective degradation. The CBM2 (Din et al. 1991) and CBM44 (Gourlay et al. 2012) exert such disruptive role on the particular substrate.

Adhesion: CBMs help the enzymes to tightly adhere to the surface of carbohydrate substrates (Montanier et al. 2009).

The studies showed that eliminating the CBM from GHs greatly reduces the enzymatic activities (Shoseyov et al. 2006). Instead, replacing the endo-β-1,4-glucanase CBM from *Bacillus subtilis* with the exoglucanase I CBM from *Trichoderma viride* resulted in higher substrate-binding affinity and also enhanced the hydrolytic activity towards the microcrystalline cellulose (Kim et al. 1998). Similar findings were also reported by Limon et al. (2001), where substituting the CBM of cellobiohydrolase II derived from *Trichoderma reesei*, with CBM of Chitinase from *Trichoderma harzianum* improved the catalytic efficiency of the enzyme.

2.5 Linkers

The catalytic and carbohydrate-binding domains of beta-glucanases are connected by short amino acid sequences called linkers. The presence of these interconnected regions was confirmed through proteolytic cleavage of enzymes. It forms flexible, extended hinges in between main domains of multidomain proteins. Linkers are different in their size and amino acid composition (Fig. 2.7) (Ruiz et al. 2016). In

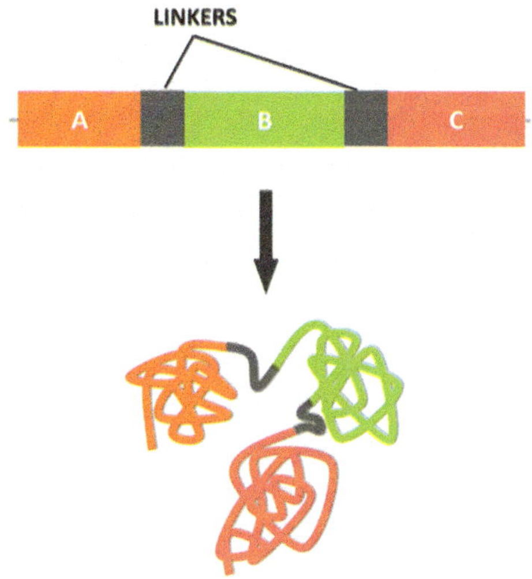

Fig. 2.7 Schematic representation of linker sequences connecting the adjacent domains

Cellulomonas fimi endoglucanases and *Trichoderma reesei* exoglucanases, the primary cleavage sites are the region located between the catalytic domain and CBM domain. This region is rich in proline or hydroxy amino acids (Gilkes et al. 1988; Tomme et al. 1988; Van Tilbeurgh et al. 1986). The length of the inter-domain linkers is varied in between 6 and 59 amino acids. Like catalytic and CBM domains, linkers of the same organisms show some sequence identity (Table 2.1).

The rigidity of the linker region is determined by proline residues. It also allows stretched conformations of enzymes (Receveur et al. 2002). The further flexibility provided by glycine residues also helps to maintain the orientation of domains (Sammond et al. 2012). In addition, the glycosylation of serine and threonine residues in the linker region is significant for preventing proteolysis, maintaining protein flexibility and conformations through maintaining the proper distance between nearby modules, as well as improving the substrate-binding affinity of enzymes (Beckham et al. 2010). Some linkers beta-glucanases are rich in glutamine, glutamate, arginine and other hydrophobic amino acids (Wootton and Drummond 1989). However, some enzymes with distinct domains do not encompass detectable linker sequences. For example, the linkers in endoglucanase (CenA) *Cellulomonas fimi* have been deleted, however, the CBM and catalytic domains of the enzymes are fully functional and structurally comparable to those of wild-type enzymes (Ting et al. 2009).

2.6 Classification of Beta-Glucanase

Beta-glucanases are carbohydrate-active enzymes classified under glycosyl/glycoside hydrolases and updated in Carbohydrate-Active enZYmes (CAZyme) Database based on genomic, biochemical and structural information. Each family in the

Table 2.1 Amino acid sequences of linker regions in beta-glucanases

Organism	Enzyme	Linker Sequence	Reference
Bacillus strain N4	Endoglucanase	TTPPSDPTPPSDPDGEPGPDPGEPDPTPPSDP	Fukumori et al. (1986)
Trichoderma reesei	Exoglucanase	PGGNRGTTTTRRPATTTGSSPGPTQS	Fägerstam et al. (1984)
Xanthomonas campestris	Endoglucanase	TTPTPTPTPTPTPTPTPTPTPTPT	Gough et al. (1990)
Streptomyces sp. strain KSM-9	Endoglucanase	PRTTPTPTP	Nakai et al. (1988)
Pseudomonas fluorescence subsp. *cellulosa*	Endoglucanase	SSAPSSVASSSSSSSVVSSTPRSSSSSVSSSVPGTSSSSSSS	Gilbert et al. (1990)
Microbispora bispora	Endoglucanase	PPTYSPSPTPSTPSPSPSQSDPGSPSPSPS	Yablonsky and Bartley (1988)
Fibrobacter succinogenes	Endoglucanase	PVSSSDMSPTSSDAVIDPTSSSAAVVDPST	McGavin et al. (1989)
Erwinia chrysanthemi	Endoglucanase	TTDPSTDTTMTPPLTNRPQPT	Guiseppi et al. (1988)
Clostridium thermocellum	Endoglucanase	PTPTPTWTSTPPSSSP	Yague et al. (1990)
Cellulomonas fimi	Exoglucanase	PTPTPTTPTPTPTTPTPTPTS	O'Neill et al. (1986)

database describes enzymes with structural similarity in the catalytic domain and CBM and also its capability to modify, degrade or create glycosidic bonds.

GH Family 1:	**Exo-β-1,4-glucanase (EC 3.2.1.74)**
	Removes sequential glucose units by cleaving (1–4)-glycosidic linkages in beta-1,4-glucans.
	Synonyms: exocellulase; exo-1,4-beta-glucosidase; exo-beta-1,4-glucosidase; beta-1,4-beta-glucanase; beta-glucosidase; 1,4-beta-D-glucan glucohydrolase.
GH Family 3:	**Endo-1,4-β-D-glucanase (EC 3.2.1.4)**
	It has endohydrolysis action on 1,4-beta-glucosidic bonds in mixed linkage beta-glucans, such as lichenin, cellulose and cereal beta-glucans.
	Synonyms: beta-1,4-glucanase; cellulase; beta-1,4-endoglucan hydrolase; cellulase A; endoglucanase D; avicelase; 1,4-(1,3;1,4)-beta-D-glucan 4-glucanohydrolase.
	Endo-1,3(4)-β-glucanase (EC 3.2.1.6)
	cleaves internal 1,3 or 1-4-linkages in beta-glucans such as laminarin, lichenin and cereal beta-glucans.
	Synonyms: laminarinase; laminaranase; endo-1,3-beta-glucanase; beta-1,3-glucanase; beta-1,3-1,4-glucanase; endo-beta-1,3(4)-glucanase; endo-beta-1,3-1,4-glucanase; 1,3-(1,3;1,4)-beta-D-glucan 3(4)-glucanohydrolase.
	Endo-1,3-β-glucanase (EC 3.2.1.39)
	cleaves 1,3-beta glucosidic bonds in 1–3 beta-glucans.
	Synonyms: glucan endo-1,3-beta-glucosidase, laminarinase; laminaranase; oligo-1,3-glucosidase; endo-1,3-β-glucanase; callase; β-1,3-glucanase; endo-(1,3)-beta-D-glucanase; kitalase; 1,3-beta-glucan 3-glucanohydrolase; endo-1,3-β-D-glucanase; endo-1,3-β-glucosidase; 1,3-β-D-glucan glucanohydrolase.
	Licheninase (EC 3.2.1.73)
	Breaks down 1,4-beta-glucosidic linkages in 1,3-1,4-beta-glucans such as lichenin and cereal beta-glucan.
	Synonyms: lichenase; 1,3;1,4-beta-glucan endohydrolase; beta-1,4-D-glucan 4-glucanohydrolase; 1,3;1,4-beta-glucan 4-glucanohydrolase.
GH Family 5:	**Cellulose 1,4-β-cellobiosidase (EC 3.2.1.91)**
	Hydrolysis of 1,4-beta-D-glucosidic linkages and releases cellobiose from the non-reducing ends of cellulose and cellotetraose.
	Synonyms: exo-cellobiohydrolase; beta-1,4-glucan cellobiohydrolase; beta-1,4-glucan cellobiohydrolase; exoglucanase; CBH 1; 1,4-beta-glucan cellobiosidase; avicelase; C1 cellulase; cellobiohydrolase I; cellobiohydrolase; exo-beta-1,4-glucan cellobiohydrolase; 1,4-beta-D-glucan cellobiohydrolase; cellobiosidase.
	Exo beta-1,3-glucanase (EC 3.2.1.58)
	Removes beta-D-glucose units from non-reducing ends of 1,3-beta-D-glucans.
	Synonyms: exo-1,3-beta-glucosidase; exo 1,3-glucanohydrolase; beta-1,3-glucan exo-hydrolase; 1,3-beta-glucan glucohydrolase glucan 1,3-beta-glucosidase.
GH Family 6:	**Cellulose 1,4-beta-cellobiosidase (EC 3.2.1.176)**

(continued)

	Hydrolyses 1,4-beta-D-glucosidic linkages and releases cellobiose from the reducing ends of cellulose.
	Synonyms: CelS; CelSS; cellulase SS; cellobiohydrolase CelS; endoglucanase SS; Cel48A.
GH Family 30:	**Endo-1,6-beta-glucanase (EC 3.2.1.75)**
	Cleaves 1,6-linkages in 1,6-beta-glucans such as lutean and pustulan Synonyms: beta-1,6-glucan hydrolase; glucan endo-1,6-beta-glucosidase; beta-1,6-beta-D-glucanase; beta-1,6-glucan 6-glucanohydrolase; beta-1,6- glucanase-pustulanase; 1,6-beta-D-glucan glucanohydrolase.
GH family 162:	**Endo-1,2-beta-glucanase (EC 3.2.1.71).**
	Randomly cleaves 1,2-glucosidic linkages in 1,2-beta-glucans
	Synonyms: glucan endo-1,2-beta-glucosidase; beta-D-1,2-glucanase; 1,2-beta-D-glucan glucanohydrolase endo-1,2-beta-D-glucanase.

2.6.1 Endo-β-1,3-Glucanase

Endo-β-1,3-glucanase, a non-cellulosic enzyme that hydrolyses internal linkages of β-1,3-glucan, is also termed laminarinase. Chesters and Bull (1963) first used the term laminarinase because the enzyme efficiently degraded laminarin (β-1,3-glucan) polysaccharides, mostly present microalgae and seaweeds. Laminarin from *Laminaria digitata* contains linear 1–3 and 1–6 glycosidic linkages. Laminarinase is a ubiquitous enzyme found in fungi, bacteria, yeast, algae, invertebrates such as earthworms and higher plants. It is associated with fungi for development, defence, β-glucan mobilization and fungal-plant pathogen interactions. Laminarinase in microorganisms assists the extracellular degradation of plant debris and in animals, it improves the digestive metabolism (Lee et al. 2014). Commonly used β-1,3-glucans for the isolation and screening of endo-β-1,3-glucanase are pachyman from mushroom *Poria cocos*, laminarin from the marine macro-alga, callose from plants, zymosan or schizophyllan from yeasts and fungi cell walls and curdlan from bacteria *Alcaligenes faecalis* var. *Myxogenes* (Fuchs et al. 2003).

Like other GHs, endo-β-1,3-glucanase also has a multi-domain structural configuration, consisting of the catalytic domain, CBM and linkers regions. Mostly bacterial laminarinases are classified in GH family 16 and characterized with a β-jelly roll folding in open cleft like catalytic domain. Two glutamate residues that occupy the inactive site region participate in typical GHs catalysis as mentioned earlier through a double displacement mechanism. Jeng et al. (2011) reported the *Thermotoga maritima* laminarinase catalytic domain (TmLamCD) has loop controlled the opening at the active site of laminarinase. It suggested that the open gateway permits the easy passage of substrates and products, that is the enzyme binds the substrates like an endo mode, whereas the closed gate allowed the efficient cleavage of polysaccharides (Fig. 2.8). The study of the catalytic domain structure of

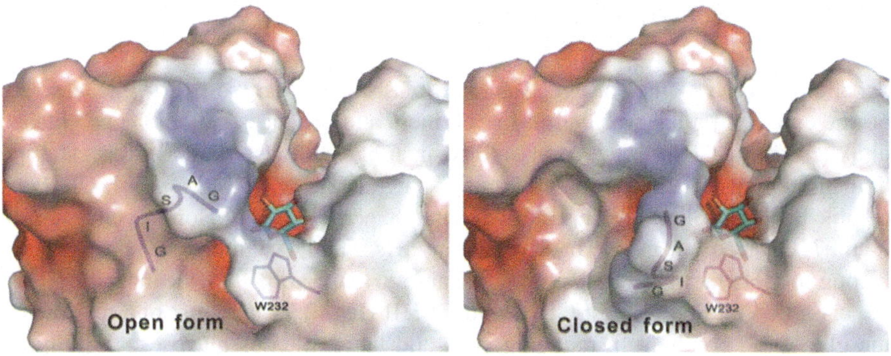

Fig. 2.8 The cleft like active site of endo-β-1,3-glucanase showing open and closed conformations (Jeng et al. 2011)

GH16 family endo-β-1,3-glucanase from *Streptomyces sioyaensis* revealed it was closed at one end by protruding two key amino acid residues, Tyr38 and Tyr134 (Hong et al. 2008).

Hong et al. (2002) isolated a β-1,3-glucanase gene from *Streptomyces sioyaensis* using an activity plate assay with pachyman as substrate. The protein had an N-terminal GH 16 catalytic domain with the ability to cleave 1,3-β-linked glucans through endo mode of action, and also had a C-terminal CBM domain of family 6. The binding assay indicated CBM binds to insoluble β-glucans. The study with the entire protein and the CBM-truncated proteins show CBM at the C-terminal end enhances S. sioyaensis β-1,3-glucanase-mediated hydrolysis of insoluble 1,3-β-glucans.

Homology modelling studies of laminarinase isolated from *Flavobacterium* sp. strain UMI-01 showed a hydrogen bonding interaction of Trp113 and Asn33 together with Phe212 at the active site region to degrade β-1,3-linked laminarin (Qin et al. 2017). Lee et al. (2014) purified an endo-beta-1,3-glucanase (laminarinases) with a molecular weight of 75 kDa which has optimum activity at 75 °C and pH 4–7. It efficiently hydrolysed laminarin into laminaribiose and glucose. According to Guan et al. (2018), the codon-optimized laminarinase gene from *Streptomyces* was expressed in the *E. coli* system yielded 19.1 U.mL^{-1} of enzymes which depolymerize curdlan polysaccharides into pentasaccharides. Molecular characterization of an endo-β-1,3-glucanase gene (bglS27) isolated from *Streptomyces* sp. S27 expressed in *E. coli* BL21 (DE3) produced the 42.7 kDa protein with GH16 family catalytic domain, family 13 CBM and a glycine-rich linker region. It preferably hydrolysed β-1,3-linked laminarin, showing 236.0 U.mg-1 of specific activity at optimum conditions of 65 °C and pH 5.5 (Shi et al. 2010).

2.6.2 Exo-Beta-1,4-Glucanase

A cellulosic type β-glucanase, cellobiohydrolase (CBH) or exo-β-1,4-glucanase act on the ends of β-glucan or cellulose polymers and produces cellobiose or glucose units. Different CBHs have been studied in fungi and bacteria and their catalytic domain belongs to different GHs families such as 5, 6, 7, 9, 48 and 74. CBHs of aerobic bacteria belong to GHs family 6 and 48, aerobic fungi belong to GHs family 6 and 7, CBHs of anaerobic bacteria are in GHs 9 and 48 families, anaerobic fungi belong to GHs family 48 (Zhang and Zhang 2013). Exo-β-1,4-glucanase or CBHs are classified into two types: CBHI and CBHII (Fig. 2.9). The first one hydrolyses the glycosidic bonds from reducing ends and latter releases glucose or cellobiose from non-reducing ends. Cel7a (CBH I) and Cel6A (CBH II) from *Trichoderma reesei* are widely studied fungal cellobiohydrolases usually expressed in *Hypocrea jecorina* for large-scale production (Suominen et al. 1993).

Cellobiohydrolases display domain architectural structure, containing a catalytic domain, cellulose-binding modules (CBMs), cell interaction motifs, linkers, fibronectin (Fn3) type III modules, etc., which provides stabilized efficient catalysis (Annamalai et al. 2016). CBHI contains N-terminal catalytic domain and C-terminal CBM, while CBHII has N-terminal CBM and C-terminal catalytic module. The most noteworthy structural feature of CBHs is the tunnel-shaped catalytic module and the catalytic amino acids are buried inside the tunnel cavity (Zhang and Zhang 2013). Catalytic core domain structural studies of CBHI (TrCBH I) of GH7 family in *T. reesei* suggested the central motif mainly constitutes of antiparallel β-sandwich loops (Stahlberg et al. 1996). During catalysis, the polysaccharide chain slides through the tunnel and progressively releases disaccharide and cellulose units. The CBMs significantly increase the proximity of enzyme to the substrates, enhance accessibility and modify the crystalline surface of substrates. CBHs or exoglucanases effectively hydrolyse microcrystalline cellulose by shedding chains from the whole structure. According to Maki et al. (2009), the enzyme that shows significant activity on Avicel (microcrystalline cellulose) and very little towards carboxymethyl cellulose (CMC) was identified as exoglucanases (CBHs). Avicel (microcrystalline cellulose or hydrocellulose) is mostly used for detecting and quantifying exo-β-1,4- glucanase activity (Wood and Bhat 1988).

Recombinant *Saccharomyces cerevisiae* was developed by sequential incorporation of three CBH-encoding genes (*Aspergillus aculeatus* cbh1 and *T. reesei* cbh1, cbh2) with a strong Ptpi constitutive promoter. The recombinant strains resulted in

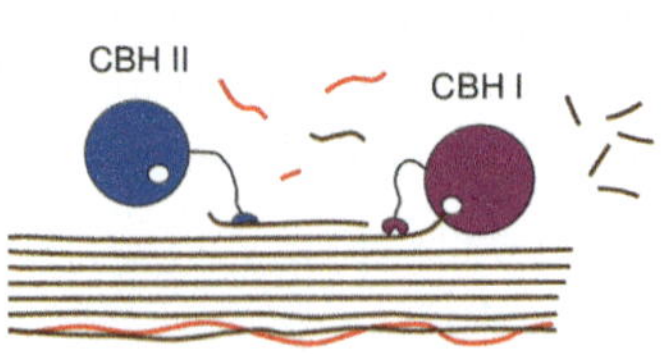

Fig. 2.9 Schematic representation of the mode of action of cellobiohydrolase I (CBH I) and II (CBH II) hydrolyse the cellulose chains sequentially from their reducing and non-reducing ends

high ethanol production than wild strains (Hong et al. 2014). Heinzelman et al. (2010) described an effectual approach based on SCHEMA recombination for screening thermostable CBHI chimaeras. Predicted 16 CBH1 chimaeras, which showed more thermostability than the parent *Talaromyces emersonii* CBHI. Sanchez et al. (2003) reported recombinant expression of a unique cellobiohydrolase Cel48C from *Paenibacillus* sp. BP-23. The reported enzyme belongs to the GH48 family with a multi-domain structure and showed maximum activity on Avicel, acid-swollen cellulose and cellodextrins at 45 °C and pH 6.0.

References

Abe K, Nakajima M, Yamashita T, Matsunaga H, Kamisuki S, Nihira T, Takahashi Y, Sugimoto N, Miyanaga A et al (2017) Biochemical and structural analyses of a bacterial endo–β1,2-glucanase reveal a new glycoside hydrolase family. J Biol Chem 292(18):7487–7506

Abou Hachem M, Nordberg Karlsson EN, Bartonek-Roxâ E, Raghothama S, Simpson PJ, Gilbert HJ, Williamson MP, Holst O (2000) Carbohydrate binding modules from a thermostable Rhodothermus marinus xylanase: cloning, expression and binding studies. Biochem J 345(1): 53–60

Agarwal PK (2006) Enzymes: an integrated view of structure, dynamics and function. Microb Cell Factories 5:2

Annamalai N, Rajeswari MV, Sivakumar N (2016) Cellobiohydrolases: role, mechanism, and recent developments. In: Gupta V (ed) Microbial enzymes in bioconversions of biomass. Biofuel and biorefinery technologies, vol 3. Springer, Cham, pp 29–35

Ardèvol A, Rovira C (2015) Reaction mechanisms in carbohydrate-active enzymes: glycoside hydrolases and glycosyltransferases. Insights from ab initio quantum mechanics/molecular mechanics dynamic simulations. J Am Chem Soc 137(24):7528–7547

Armenta S, Moreno-Mendieta S, Sánchez-Cuapio Z, Sánchez S, Rodríguez-Sanoja R (2017) Advances in molecular engineering of carbohydrate-binding modules. Proteins 85(9): 1602–1617

Baroroh U, Yusuf M, Rachman SD, Ishmayana S, Syamsunarno M, Levita J, Subroto T (2017) The importance of surface-binding site towards starch-Adsorptivity level in α-amylase: a review on structural point of view. Enzyme research 2017:4086845

Beckham GT, Bomble YJ, Matthews JF, Taylor CB, Resch MG, Yarbrough JM, Decker SR, Bu L, Zhao X et al (2010) The O-glycosylated linker from the Trichoderma reesei Family 7 cellulase is a flexible, disordered protein. Biophys J 99(11):3773–3781

Boraston AB, Kwan E, Chiu P, Warren AJ, Kilburn DG (2002) Recognition and hydrolysis of non-crystalline cellulose. J Biol Chem 278:6120–6127

Cantarel BL, Coutinho PM, Rancurel C, Bernard T, Lombard V, Henrissat B (2009) The carbohydrate-active EnZymes database (CAZy): an expert resource for Glycogenomics. En. Nucleic Acids Res 37:D233–D238

Chesters CG, Bull AT (1963) The enzymatic degradation of laminarin. 2. The multicomponent nature of fungal laminarinases. Biochem J 86(1):31–38

Codera V, Gilbert HJ, Faijes M, Planas A (2015) Carbohydrate-binding module assisting glycosynthase-catalysed polymerizations. Biochem J 470(1):15–22

Datta S, Sapra R (2011) Cellulases and hemicellulases for biomass degradation: an introduction. In: Simmons B (ed) Chemical and biochemical catalysis for next generation biofuels. Royal Society of Chemistry, pp 115–135

Davies G, Henrissat B (1995) Structures and mechanisms of glycosyl hydrolases. Structure 3(9): 853–859

Din N, Gilkes NR, Tekant B, Miller RC, Warren RAJ, Kilburn DG (1991) Non–hydrolytic disruption of cellulose fibres by the binding domain of a bacterial cellulase. Nat Biotechnol 9(11):1096–1099

Espinosa JF, Montero E, Vian A, Garcia JL, Dietrich H, Schmidt RR, Martín-Lomas M et al (1998) For an example where O- and C- glycosides bind in different conformations with respect to the intersaccharide torsions. J Am Chem Soc 120:1309–1318

Fägerstam LG, Göran Pettersson L, Åke Engström J (1984) The primary structure of a 1,4-,3-glucan cellobiohydrolase from the fungus Trichoderma reesei QM9414. FEBS Lett 167(2):309–315

Fuchs KP, Zverlov VV, Velikodvorskaya GA, Lottspeich F, Schwarz WH (2003) Lic16A of clostridium thermocellum, a non-cellulosomal, highly complex endo-b-1,3-glucanase bound to the outer cell surface. Microbiology (Reading) 149(4):1021–1031

Fukumori F, Kudo T, Narahashi Y, Horikoshi K (1986) Molecular cloning and nucleotide sequence of the alkaline cellulase gene from the alkalophilic bacillus sp. strain 1139. J Gen Microbiol 132(8):2329–2335

Fulop L, Ponyi T (2015) Classification of glycosyl hydrolases based on structural homology. Journal of Universal Science 2(1):1–9

Gilbert HJ, Hall J, Hazlewood GP, Ferreira LMA (1990) The N-terminal region of an endoglucanase from Pseudomonas fluorescens subspecies cellulosa constitutes a cellulose-binding domain that is distinct from the catalytic Centre. Mol Microbiol 4(5):759–767

Gilkes NR, Warren RAJ, Miller RC Jr, Kilburn DG (1988) Precise excision of the cellulose binding domains from two Cellulomonas fimi cellulases by a homologous protease and the effect on catalysis. J Biol Chem 263(21):10401–10407

Gough CL, Dow JM, Keen J, Henrissat B, Daniels MJ (1990) Nucleotide sequence of the gene encoding the major endoglucanase of Xanthomonas campestris pv. Campestris Gene 89(1): 53–59

Gourlay K, Arantes V, Saddler JN (2012) Use of substructure-specific carbohydrate binding modules to track changes in cellulose accessibility and surface morphology during the amorphogenesis step of enzymatic hydrolysis. Biotechnol Biofuels 5(1):51

Guan N, Yu J, Liu W, Chen R (2018) A high-yielding enzymatic process for production of bioactive β-1,3-glucan oligosaccharides with defined size. J Glycobiol 07(1):131

Guiseppi A, Cami B, Aymeric JL, Ball G, Creuzet N (1988) Homology between endoglucanase Z of Erwinia chrysanthemi and endoglucanases of Bacillus subtilis and alkalophilic bacillus. Mol Microbiol 2(1):159–164

Heinzelman P, Komor R, Kanaan A, Romero P, Yu X, Mohler S, Snow C, Arnold F (2010) Efficient screening of fungal cellobiohydrolase class I enzymes for thermostabilizing sequence blocks by SCHEMA structure-guided recombination. Protein Eng Des Sel 23(11):871–880

Herve C, Rogowski A, Blake AW, Marcus SE, Gilbert HJ, Knox JP (2010) Carbohydrate-binding modules promote the enzymatic deconstruction of intact plant cell walls by targeting and proximity effects. Proc Natl Acad Sci U S A 107(34):15293–15298

Hong TY, Cheng CW, Huang JW, Meng M (2002) Isolation and biochemical characterization of an endo-1,3-β-glucanase from Streptomyces sioyaensis containing a C-terminal family 6 carbohydrate binding module that binds to 1,3-β-glucan. Microbiology 148:1151–1159

Hong TY, Hsiao YY, Meng M, Li TT (2008) The 1.5 a structure of endo-1,3-beta-glucanase from Streptomyces sioyaensis: evolution of the active-site structure for 1,3-beta-glucan-binding specificity and hydrolysis. Acta Crystallogr D Biol Crystallogr 64(9):964–970

Hong J, Yang H, Zhang K, Liu C, Zou S, Zhang M (2014) Development of a cellulolytic Saccharomyces cerevisiae strain with enhanced cellobiohydrolase activity. World J Microbiol Biotechnol 30(11):2985–2993

Jeng WY, Wang NC, Lin CT, Shyur LF, Wang AH (2011) Crystal structures of the laminarinase catalytic domain from Thermotoga maritima MSB8 in complex with inhibitors: essential residues for β-1,3- and β-1,4-glucan selection. J Biol Chem 286(52):45030–45040

Kim H, Goto M, Jeong HJ, Jung KH, Kwon I, Furukawa K (1998) Functional analysis of a hybrid endoglucanase of bacterial origin having a cellulose binding domain from a fungal exoglucanase. Appl Biochem Biotechnol 75(2–3):193–204

Laskowski RA, Luscombe NM, Swindells MB, Thornton JM (1996) Protein clefts in molecular recognition and function. Protein Sci 5(12):2438–2452

Lee K, Arai T, Ibrahim D, Kosugi A, Prawitwong P, Lan D, Murata Y, Mori Y (2014) Purification and characterization of a thermostable laminarinase from *Penicillium rolfsii* c3-2(1) IBRL. Bioresources 9(1):1072–1084

Limon MC, Margolles-Clark E, Benítez T, Penttilä M (2001) Addition of substrate-binding domains increases substrate-binding capacity and specific activity of a chitinase from Trichoderma harzianum. FEMS Microbiol Lett 198(1):57–63

Maki M, Leung KT, Qin W (2009) The prospects of cellulase-producing bacteria for the bioconversion of lignocellulosic biomass. Int J Biol Sci 5(5):500–516

McGavin MJ, Forsberg CW, Crosby B, Bell AW, Dignard D, Thomas DY (1989) Structure of the cel-3 gene from Fibrobacter succinogenes S85 and characteristics of the encoded gene product, endoglucanase 3. J Bacteriol 171(10):5587–5595

McLean BW, Boraston AB, Brouwer D, Sanaie N, Fyfe CA, Warren RA, Kilburn DG, Haynes CA (2002) Carbohydrate-binding modules recognize fine substructures of cellulose. J Biol Chem 277(52):50245–50254

Montanier C, van Bueren AL, Dumon C, Flint JE, Correia MA, Prates JA, Firbank SJ, Lewis RJ, Grondin GG et al (2009) Evidence that family 35 carbohydrate binding modules display conserved specificity but divergent function. Proc Natl Acad Sci U S A 106(9):3065–3070

Nakai R, Horinouchi S, Beppu T (1988) Cloning and nucleotide sequence of a cellulase gene casA, from an alkalophilic Streptomyces strain. Gene 65(2):229–238

Nishimura T, Bignon C, Allouch J, Czjzek M, Darbon H, Watanabe T, Henrissat B (2001) *Streptomyces matensis* laminaripentose hydrolase is an 'inverting' beta-1,3-glucanase. FEBS Lett 499(1–2):187–190

O'Neill GP, Goh SH, Warren RAJ, Kilburn DG, Miller RC Jr (1986) Structure of the gene encoding the exoglucanase of Cellulomonas fimi. Gene 44(2–3):325–330

Planas A (2000) Bacterial 1,3-1,4-beta-glucanases: structure, function and protein engineering. Biochim Biophys Acta 1543(2):361–382

Qin HM, Miyakawa T, Inoue A, Nakamura A, Nishiyama R, Ojima T, Tanokura M (2017) Laminarinase from *Flavobacterium* sp. reveals the structural basis of thermostability and substrate specificity. Sci Rep 7(1):11425

Receveur V, Czjzek M, Schülein M, Panine P, Henrissat B (2002) Dimension, shape, and conformational flexibility of a two domain fungal cellulase in solution probed by small angle X-ray scattering. J Biol Chem 277(43):40887–40892

Ruiz DM, Turowski VR, Murakami MT (2016) Effects of the linker region on the structure and function of modular GH5 cellulases. Sci Rep 6:28504

Sajith S, Priji P, Sreedevi S, Benjamin S (2016) An overview on fungal cellulases with an industrial perspective. J Nutr Food Sci 6:461

Sammond DW, Payne CM, Brunecky R, Himmel ME, Crowley MF, Beckham GT (2012) Cellulase linkers are optimized based on domain type and function: insights from sequence analysis, biophysical measurements, and molecular simulation. PLoS One 7(11):e48615

Sanchez MM, Pastor FI, Diaz P (2003) Exo-mode of action of cellobiohydrolase Cel48C from Paenibacillus sp. BP-23. A unique type of cellulase among Bacillales. Eur J Biochem 270(13): 2913–2919

Sathya TA, Khan M (2014) Diversity of glycosyl hydrolase enzymes from metagenome and their application in food industry. J Food Sci 79(11):R2149–R2156

Shi P, Yao G, Yang P, Li N, Luo H, Bai Y, Wang Y, Yao B (2010) Cloning, characterization, and antifungal activity of an endo-1,3-beta-D: -glucanase from Streptomyces sp. S27. Appl Microbiol Biotechnol 85(5):1483–1490

Shoseyov O, Shani Z, Levy I (2006) Carbohydrate binding modules: biochemical properties and novel applications. Microbiol Mol Biol Rev 70(2):283–295

Stahlberg J, Divne C, Koivula A, Piens K, Claeyssens M, Teeri TT, Jones TA (1996) Activity studies and crystal structures of catalytically deficient mutants of cellobiohydrolase I from Trichoderma reesei. J Mol Biol 264(2):337–349

Stone BA, Clarke AE (1992) Chemistry and biology of 1,3-LGlucans: Bundoora. La Trobe University Press, Australia

Suominen PL, Mäntylä AL, Karhunen T, Hakola S, Nevalainen H (1993) High frequency one-step gene replacement in Trichoderma reesei. II. Effects of deletions of individual cellulase genes. Mol Gen Genet 241(5–6):523–530

Teegardin KA, James S, Barabote RD (2017) Bioinformatic analysis of glycoside hydrolases in the proteomes of mesophilic and thermophilic Actinobacteria. MOJ Proteom Bioinformatics 5(3): 75–81

Ting CL, Makarov DE, Wang ZG (2009) A kinetic model for the enzymatic action of cellulase. J Phys Chem B 113(14):4970–4977

Tomme P, Van Tilbeurgh H, Pettersson G, Van Damme J, Vandekerckhove J, Knowles J, Teeri T, Claeyssens M (1988) Studies of the cellulolytic system of Trichoderma reesei QM 9414. Analysis of domain function in two cellobiohydrolases by limited proteolysis. Eur J Biochem 170(3):575–581

Van Tilbeurgh H, Tomme P, Claeyssens M, Bhikhabhai R, Pettersson G (1986) Limited proteolysis of the cellobiohydrolase I from Trichoderma reesei. FEBS Lett 204(2):223–227

White WB, Bird HR, Sunde ML, Marlett JA, Prentice NA, Burger WC (1983) Viscosity of β-D-glucan as a factor in the enzymatic improvement of barley for chicks. Poult Sci 62(5):853–862

Wood TM, Bhat KM (1988) Methods for measuring cellulase activities. Methods Enzymol 160:87–112

Wootton JC, Drummond MH (1989) The Q-linker: a class of interdomain sequences found in bacterial multidomain regulatory proteins. Protein Eng 2(7):535–543

Yablonsky MD, Bartley TKO (1988) Characterization and cloning of the cellulase complex of Microbispora bispora. Elliston, S. K. Kahrs, Z. P. Shalita, and D. E FEMS Symp: Eveleigh 43: 249–266

Yague E, Béguin P, Aubert JP (1990) Nucleotide sequence and deletion analysis of the cellulase-encoding gene celH of clostridium thermocellum. Gene 89(1):61–67

Zhang XZ, Zhang YHP (2013) Cellulases: characteristics, sources, production, and applications. In: Yang ST, El-Enshasy HA, Thongchul N (eds) Bioprocessing technologies in biorefinery for sustainable production of fuels, chemicals, and polymers. Wiley, New York, pp 131–146

Chapter 3
Beta-Glucanase: Diverse Bacterial Sources and its Applications

P. Nisha

Abstract Beta-glucans are found in various microorganisms like Bacteria, Actinomycetes, Fungi, Yeast, and Algae also found in some plants such as oats and barley. These are enzymes that are capable to break down a polysaccharide beta-glucan into glucose subunits by hydrolyzing the glycosidic bond. Numerous wild and recombinant beta-glucanases are isolated and well characterized from a wide range of microorganisms. These enzymes from microbes are specific on the substrates β-glucan and capable to cleave within the chain. Microbial beta-glucanases are economically important enzymes and have applications in various industries like textile, brewing and food industries also in agriculture. This chapter deals with the advisable perceptive of the beta-glucanase enzyme from the diverse bacterial community and their applications.

Keywords Beta-glucanases · Endo-glycosidase · Polysaccharides · Beta-1,3-glucanase · Endo-1,3-beta-glucanase

3.1 Introduction

β-Glucanases are enzymes with a group of carbohydrates which are capable to break down glycosidic bonds within beta-glucan. Glucanases are the enzymes that have undergone breakdown of glucans, are a portion of some plant cell wall. 1,4 beta-glycosidic bonds in beta-D Glucan, cellulose, and lichenin are hydrolyzed by these glucanases also glucanases, breaks the cellulose bonds in the internal portion (Barras and Stone 1969). Exo glucanases are responsible for the cleavage of 2,4 units of cellulose polysaccharide chains at the exposed end (Tsai et al. 2015). Xyloglucanases are enzymes that have undergone the breakdowns of hemicellulose xyloglucan (Vincken et al. 1997). Glucanases are the enzymes classified according to the breakdown process of polysaccharides. Endo-1,3-beta-glucanases are involved in the hydrolysis of 1,3-β-glucosyl linkages seen internally. Beta-1,3

P. Nisha (✉)
Department of Biosciences, MES College, Ernakulam, Kerala, India

N. S. Pradeep, L. K. Edison (eds.), *Microbial Beta Glucanases*, Interdisciplinary Biotechnological Advances, https://doi.org/10.1007/978-981-19-6466-4_3

glucanases are well-advised as more supportive and act as an enzymatic tool for the degradation of β-1,3-glucan, which produce (1 → 3) linked β-glucan oligosaccharides. The glycoside hydrolase β-1,3-glucanases are found in bacteria, plants, and fungi mainly as cell wall components, which work on 1,3 beta-glycosidic bonds of l beta-1,3-glucans and transfer or hydrolyze the glycosides (Kobayashi et al. 2016; Kusaykin et al. 2017). Beta-1,3-glucanases are divided into 2 classes based on the hydrolysis position:

1. endo-type,
2. exo-type.

This enzyme has various names:

1-3, 1-6-beta-glucan, 1-3,1-6-bêta-glucane, beta-1,3-D glucan, Beta-1,3/1,6-D Glucan, Bêta-1,3/1,6-D Glucane, Beta 1,3/1,6 Glucan, Beta 1,3/1,6 Glucane, Bêta-1,3/1,6 Glucanes, Beta-1,3/1,6 Glucans, Beta-1,3-D Glucan, Bêta-1,3-D Glucane, Bêta-1,3-D-glucane, Beta-1,3-Glucan, Bêta-1,3-Glucane, Beta 1,3 Glucan, Bêta 1,3 Glucane, Beta 1,6 Glucan, Bêta 1,6 Glucane, beta-1-6,1,3-beta-glucan, Beta 1,3 Glucans, Bêta 1,3 Glucanes, Beta Glucan, Beta-Glucan, Bêta-glucanes, Beta Glucanos, Beta-Glucans, Beta Glycans, Beta-Glycans, Grifolan (GRN), Lentinan, Beta Glucans of Oats, PGG Glucans, Poly [1-6] Beta D Glucopyranosyl [1-3] Beta D Glucopyranose, Schizophyllan, Yeast derived Beta glucans, Beta Glucane d'Avoine, Barley Beta-Glucan, Beta Glucane, Beta Glucane d'Avoine, Beta Glucane d'Orge, Bêta-Glucane Dérivé de la Levure.

3.2 Diverse Bacterial Sources

Beta 1,3-Glucanase from diverse microbes is used as an important device for the degradation of beta 13 glucan and forms (1 → 3)-linked beta-glucan oligosaccharides. Beta 1,3 glucanase producing microorganisms which isolated from several environments with rich sources of beta-1,3-glucan-degrading microorganisms present in soil of *Wolfiporia extensa*, and subjected to sequencing. The results of sequencing observed the presence of various genera of *Arthrobacter sp.* of 0.78%, *Streptomyces* sp. of 1.90%, the order of Actinomycetales was around 8.64% and phylum Actinobacteria was about 18.64% were determined in soil with *Poria cocos* cultivation also for the isolation of microorganisms with glucan-degrading ability, Actinomycetes are also considered as the glucan-degrading ability candidates. Only 11 microbial members show beta 1,3-glucan-degrading activity out of 58 isolates. The microbial strain SYBC17 reveals the highest yield of 1.02 U/mg and the bacterial isolate SYBCQL belongs to the genus Kitasatospora. The Two genes 17 Q and 17 W, which were encoding an enzyme β-1,3 glucanase in SYBC17 and another gene, QLK1 in SYBCQL subjected to cloning and also expression for its molecular level verification, to check the contribution of beta 1,3 glucanase activity to degrade beta 1,3 glucan. Final findings of the study collectively showed the ability of microbial isolates to secrete β 1,3 glucanase obtained supporting with sequencing, genes expression results, and molecular analysis (Wu et al. 2018).

Beta-glucanase enzymes are present in microbes and plants, and they belong to different families of glucanases, which also shows indifference in the sequences, no relationship within three-dimensional structures. The plant glucanases are in the family of GH Family 17 which follow a (Beta/Alpha) 8 TIM barrel fold, whereas the bacterial beta-glucanases are under GH Family 16 having beta sandwich folds. These two glucanases from bacteria and plant sources show some of the similarities in substrate specificity and their independent activity while its structural differences which give functional activity independently despite their structural differences give another example of functional evolution (Doolittle 1994). There are some other numerous enzymes other than glucanases, which can catalyze independently in a similar reaction. For example, Ser/His/Asp is a catalytic triad found in various protein folds of minimum and is not to be considered as homologous in structure (Dodson and Wlodawer 1998; Håkansson et al. 2000).

Pyrococcus furiosus laminarinase, Lam A—PF0076 an endo-glycosidase which can be able to hydrolyze β 1,3 gluco oligosaccharides, has specificity on saccharides by using 4 methyl umbelliferyl β glucosides with other linkages (Van Lieshout et al. 2004). The carbohydrate-binding site of 1,3 1,4 beta-D glucan 4 glucanohydrolase isolated *Bacillus licheniformis* has explored with synthetic 4 methyl umbelliferyl beta-D-glucan oligosaccharides and is retaining the endo-glycosidase with a binding site of four glucopyranosyl binding subunits (Malet and Planas 1997). The unknown activities of 1,3 1,4 beta endoglucanase were identified in bacterial species like *E. faecium, Streptococcus* sp., *Bacteroides,* and *Clostridium* sp. capable of degradation of diversely linked beta-glucans in chicken's small intestine and *Streptococcus bovis* with the 1,3 1,4 beta endoglucanase activity. Strict anaerobes like *Bacteroides uniformis, B. ovatus*, presumably *Clostridium perfringens* and also *B. capillosus* were identified (Beckmann et al. 2006).

Enterococcus faecium present in the intestine of broiler chicken with an endo 1,3 1,4 endo beta-glucanase, lichenase which was catalyzed on 1,3 1,4 beta-glucans had linked as mixed, and were determined by using the method of reducing sugar assay (Beckmann et al. 2000). Numerous bacterial recombinant and wild enzymes are characterized from several bacterial strains, like *Paenibacillus* sp. (Hong and Meng 2003), *Streptomyces* (Hong et al. 2002), *Cellulosimicrobium cellulans* (Tanabe and Oda 2011), *Arthrobacter* sp. (Pang et al. 2005; Okazaki et al. 2007), *Bacillus circulans* (Asano et al. 2014), *Nocardiopsis* sp. (Masuda et al. 2006), and *Thermotoga neapolitana (*Zverlov et al. 2001). A β-1,3-glucanase (LpGluA) purified from culture broth of *Laceyella putida* JAM FM3001 isolated from deep subseafloor and estimated the molecular mass of the enzyme around 36 kDa (Kobayashi et al. 2016).

1,3beta-glucanase isolated, purified from *Paenibacillus* sp. with a molecular weight of 44 K Da, the specific activity of 28-fold, catalyzes the formation of 1,3-beta-linkage by the hydrolysis of glucans (Hong and Meng 2003). Beta 1,3 glucanase gene which was cloned from *Paenibacillus* sp., encoding a protein with the presence of 1793 amino acids, and large protein, designated as LamA, with many putative functional units (Cheng et al. 2009). The catalytic domain structure of 1,3 beta-glucanase composed of 278 amino acids, isolated from *Streptomyces sioyaensis*, a family 16 (GHF16) member of glycosyl hydrolase, has specifically breakdown the 1,3 beta-glucan linked with glucan substrate having glycosidic bond (Hong et al. 2008). An extracellular alkaline stable novel beta 1,3 glucanase isolated from *Bacillus clausii* NM-1 and purified with ammonium sulfate precipitation (Miyanishi et al. 2003). A novel beta 1,3 glucanase gene names *lam C* was cloned from *Corallococcus* sp. EGB, with the occurrence of a glycoside hydrolase family 16 GH16 catalytic module and fascin-like module and LamC demonstrates the hydrolyzing activity towards the several polysaccharides broadly (Zhou et al. 2017). bglS27 gene, endo 1,3 beta-D glucanase, was expressed in *E.coli* BL21 strains, which cloned from *Streptomyces* sp. strain S27, the gene contains 1362 bp and encodes a protein of 453 amino acids with a calculated molecular mass of 42.7 kDa (Shi et al. 2010).

Purified and crystallized (hanging drop vapor diffusion) endo 1,3 beta-glucanase from *Arthrobacter* sp. Which belonged to the group *P*41, with unit cell parameters $a = 71.31$, $c = 60.07$ Å, and having one molecule/asymmetric unit (Pang et al. 2005). The classification details of the phylum taxa level represented some of the phyla such as Bacteroidetes—5.95%, Acidobacteria—12.41%, Actinobacteria—18.64%, Proteobacteria—42.08% which were isolated from soil samples from Poria cocos cultivation soil (Wu et al. 2018). *Bacteroides* species from the human gut, basically involved in reactions related to dietary beta 1,3 glucans were studied in detail on the structural, biochemical analysis of proteins; results of the above studies revealed the presence of unpredicted diversity in PUL glycosidases and glycan-binding proteins in its substrate specificity to beta 1,3 glucan linkage and also the branching patterns of the same (Déjean et al. 2020).

Bacillus uniformis JCM 13288 T is capable of maintaining the human gut microbial homeostasis, which found to grow on pustulan, laminarin, and porphyran and can share glycans by utilizing beta-glucans in human gut bacteria. Bacterial genome has 5.05 megabase pairs and 4868 coding genes of proteins. From the genome sequence, two loci for β-glucans with putative polysaccharide utilization were identified also cloned, expressed, purified, and characterized the four putative associated genes. *B. uniformis* JCM 13288 having three glycoside hydrolases enzymes (GHs), *Bu*GH16, *Bu*GH30, and *Bu* GH158 are endo-acting enzymes, can cleave linear endo or exo beta 1-3 linkages, *Bu*GH 30 which acts as an exo-acting enzyme has a key role in the digestion of beta-glucans by cleaving a long chain beta 1,6 linked glucans and got the final product gentiobiose. *Bu* GH158 have been able to generate oligosaccharides from the polymerization of soluble curdlan, gut microbes can utilize these oligosaccharides, and they can't be able to survive in the presence of laminarin. *Bacillus uniformis* strain JCM 13288 T plays an important role in maintaining homeostasis in gut microbiota because this bacterial species have the ability to utilize beta-glucans as well as share glycan with microbes which are active in the gut (Singh et al. 2020). The structural and functional characteristics of various enzymes, hemicellulase, cellulase, β 1-3 glucanase and endo β 1,4 glucanase were studied, here cellulose like beta 1,4 glucans digested by endo β-1,4 glucanase enzyme, which included in glycosyl hydrolase family. It slowly hydrolyzes crystalline cellulose, has an enzymatic activity of both lichenase and endo beta 1,4 glucanase enzymes and also revealed the capability of beta 1,4 glucans degradation. Structural basis for the substrate specificity of a *Bacillus sp.* 1,3-1,4 beta-glucanase studied, reaction found that the selective digestion of beta 1,4 glycosidic bond in 1,3,1,4 beta-glucanase enzyme, these glycosidic bonds was in 3O substituted glucopyranosyl units mixed linkage with polysaccharides. (Gaiser et al. 2006).

The enzymatic studies including protein engineering, X-ray crystallography and the mechanistic enzymological studies on bacterial beta with jellyroll sandwich fold of the enzyme family 16 glycosyl hydrolase glucanase were disclosed binding cleft of substrate with 6 subsites (Planas 2000). A similar study on 1,3-1,4-beta-D-glucanase which was isolated from *Fibrobacter succinogenes* also known as Fsbeta-glucanase is included in 16 glycosyl hydrolase family can able to hydrolyze beta 1,3 linkage and the bond of beta 1,4, glycoside. As a part of the continuation of this study, recombinant Fs beta-glucanase enzyme was reconstructed and analyzed the amino acid residues 1 to 258 and the catalytic domain was visualized and concluded that the enzymatic activity and thermal stability were higher when compared with the full-length enzyme. The overall topology consists of two β sheets as anti-parallel with eight-strands organized as jellyroll beta-sandwich, which was similar to the carbohydrate-binding modules and glycosyl hydrolases folds. Fs beta-glucanase is the beta-glucanase enzyme with naturally occurring reversed sequences that are structurally compared with circularly permuted *Bacillus* sp. enzyme and the outcome of the study shows 70% sequence recognition with the parent enzyme. The Fs beta-glucanase enzyme in its structural high resolution, provides various enzymatic activities by the influence of a series of mutant glucanase and the modification processes of engineered enzymes with structural stability and the enhanced enzyme activity (Planas 2000).

Fibrobacter succinogenes produces an enzyme Fs Beta-Glucanase/1,31,4, D glucanase which is classified in 16 glycosyl hydrolase and are compete to hydrolyze the glycosidic bind beta 1.4 and beta 1,3 glycosidic linkage with mixed-linked glucans. Fs Beta-glucanase structural resolution gives the elaborate structural explanations of various capabilities in its enzymatic activities from structural rationale for the various activities acquired from mutant glucanases and observed the structural stability of an enzyme and increased enzyme activity on the basis of the biological processes by engineered enzymes (Tsai et al. 2003a, b). The specific binding site of an enzyme which can ability to degrade the polysaccharides, here the few numbers of subsites are involved which contain the sugar types with residues of saccharyl units and a specific OH substitutions. The detailed architectures of the subsites may allow the substrate discrimination within the various types of glycosidic linkages, and these studies revealed the orientation features of an apolymeric substrate with its monosaccharyl residues. Every enzyme has its active site, and an enzyme glycosyl hydrolase has actively performed its hydrolysis reactions on scissile glycosidic bond in endo glucanases with the nucleophile. Generally these characteristic features of acid/base catalyst with its residues help to explain about the transition state and substrate binding stabilization features in detail (Gaiser et al. 2006). Glucanase enzymes are mainly participating in the degradation of glucans, few of the research works reported about the identification of a new endo beta 1,3 glucanase Mz186 isolated from *Mesoflavibacter zeaxanthinifaciens* S86. Mz186 and amino acid sequences which revealed 45.1% of similarity with *Leeuwenhoekiella blandensis*, which located in a glycosyl hydrolase 16 family. Recombinant and purified form of rMz186 demonstrates a maximum enzyme activity at factors of pH 8, and at 50° C against the laminarin, this enzyme was more activated, i.e., 187% only in the presence of 2.5 mM manganese also more stable at 50° C for 1 hr. This, recombinant rMz186 has the substrate specific activities against various enzymes Beta-glucan. 261unit/mg, lichenin 115unit/mg and laminarin, 121unit/mg, degrade laminarin enzyme and laminarin oligosaccharides. The biochemical and molecular properties exposed the characteristics of rMz186 provide the distinctive features of beta 1,3 glucanase and can be used in various areas/industries like drug, agriculture, biochemical and bioethanol production, etc. as a potent candidate (Lee et al. 2014).

Laminarinase from a marine microbe *Flavobacterium* sp. strain UMI-01 is one of the new glycosyl hydrolase 16 family member and is able to undergo hydrolysis reaction of glycosidic bond of beta-glucan by degrading the beta 1,3,glycosyl linkages. To interpret the structural characteristics with thermostability, specificity on substrate was checked and found the resolution of laminarin was ULam111 at 1.60 Å in its crystalline structure. The detailed structure explained that the beta sheet of an enzyme from the catalytic cleft and a calcium binding site on its opposite site of this β sheet increase the degradation capability mainly in binding site near the substrate and its thermostability, also Cys 31 to Cys 34 is the disulfide bridge located on β_2–β_3 loop of the enzyme. In some domains laminarin linked with 1.3.1.4 glucan with some substrates and also binds with catalytic cleft in the subsite 3. Some of the binding sites of laminarin, Trp113, Asn33, etc. were combined with suitable substrates and formed a hydrogen bond and capable to breaking down of beta 1,3, laminarin. This study with structural and molecular information gives a novel visual image concerning with substrate recognition and thermostability, these details of laminarin will modify the design of an industrial biological catalyst (Qin et al. 2017).

The sequences and substrate specificity of bacterial and archeal endo beta 1.3 glucanase are different while they belonging to the family 16 were different and both are sharing the beta jelly roll fold. A group of researchers studied about the crystalline structure of an enzyme laminarinase, which was isolated from *Pyrococcus furiosus*, i.e., *pf*LamA is isolated from an extremophile/hyperthermophilic microbe, which determined the resolution at 2.1 Å. At the entering way of a catalytic cleft, the structural studies on *pf*LamA revealed the presence of 6 residues, i.e., 72–77. A deletion of mutant strain of *pf*LamA was deficient with 72–75 residues *pf*LamA, lacking residues 72–75, which are able to break down by hydrolysis of the mixed linkage of beta 1,3-1,4 glucan and lichenin than the wild-type protein (Ilari et al. 2009). A metagenomic library consisted with artificial bacterial clones of 5760 numbers were prepared with extracted DNA from large bowel microbiota BALB/c mice by using *Escherichia coli* DH10B strain as host cell. Library of beta-glucanase was noticed to be the presence of DNA inserts with 61 randomly clones consisting of average size around 55Kbp. The functional screen conducted and detected three clones with β-glucanase activity by using congo red in lichenin agar, these three cloned cells which were sequenced and composed and the open reading frame from the results which encoded with important proteins and shows similarity with glucanolytic/laminarinase and lichenases, were detected with the reference with the databases. Some of the other genes having functions including acquisition of nutrients, sensing environmental factors and co aggregation were also checked. The DNA inserted in the cloned cells formed shows similarity with the sequences of *Bacteroides* sp. Bacterial artificial chromosomal library prepared by using combination of cells from cecal and colonic portions from a BALB/c mice and the restriction endonuclease analysis results revealed that the clones consisted of size between 8 and 150Kbp, DNA fragments are unique and some of the clones with multiple DNA fragments. These restriction pattern of bacterial artificial chromosome was with multiple DNA fragments in Notl digests, the genome which equivalent to 3 M base pairs and the library having 100 bacterial genomes (Jens et al. 2005). The four beta-glucanase enzymes were phylogenetically compared and the sequences were similar to enzymes like lichenases-1,3,1,4, β-glucanase, and laminarinase-1,3 (4) β-glucanases.

A β-1,3-endoglucanase from *Streptomyces rutgersensis* purified by the methods, like fractional precipitation with ammonium sulfate, ion exchange chromatography, and hydrophobic chromatography, got 11.74-fold purification with 12.53% yield. One protein band obtained in purification by SDSPAGE of the protein showed that molecular mass of the enzyme was 41.25 kD obtained in mass spectrometry, the isoelectric point was between pH 4.2–4.4 and pH 7, and 50 °C was the optimal β-glucanase catalytic activity. An enzyme was active towards β-1,3 linkages with glucose polymers, hydrolyzed cell wall β-glucan of *Saccharomyces cerevisiae* and different molecular size β-glucans were the reaction products (Javmen et al. 2013). Bi-functional cellulase/xylanase is secreted by *Cellulomonas flavigena* CDBB-531 with 49KDa of molecular weight. The thin layer chromatographic results of degradation products confirmed the presence of glucose, cellobiose, cellotetraose, etc. when cellulase was reacted with a substrate carboxymethyl cellulose (CMC). This bifunctional enzyme has optimum pH 6 for cellulase activity, optimum temperature 50 °C. The purified protein sequences of predicted amino acid show similar catalytic activities domain of different glycosyl hydrolases of family 10 (Pérez-Avalos et al. 2008).

Endo beta 1,4 glucanase isolated from *Streptomyces* sp. G12 strain were recombinantly expressed in *E.coli*, named as rCelStrep and developed for enhanced biocatalytic activity of the hydrolysis of cellulose into small monosaccharides. Automated screening of rCelStrep which developed with the direct mutants and were applied for the generation of the library. The more active variant strains were used to screen the gene library with the existence of 10,000 random mutant genes and diversity was enclosed to the gene by using error-prone PCR method. Two thousand two hundred active clones were screened out on carboxymethyl cellulose agar plates as part of primary screening and these clones were subjected to quantitative screening towards AZO-CMC as secondary screening method. There are five tested mutants exhibiting 30% betterment in yields of bioconversion and showed more hydrolytic activity than the wild-type enzymes when compared with the same, when subjected to characterization and studied the lignocellulosic biomass conversion of *Arundo donax*. Homology models of three-dimensional structures of catalytic binding modules of rCelStrep obtained and mutation localization on these models allowed to expect the functions and structural relations of mutants (Cecchini et al. 2018).

In biorefineries, lignocellulosic biomasses hydrolyzed, processes require enzymes and mixtures which mixtures of enzymes which works on carbohydrates matrixes complex matrix comprising of various activities like, endoglucanase, xylanases, cellobiohydrolase, Beta glucosidase and arabinofuranosidase enzymatic activities, such as endoglucanase, β-glucosidase, and arabinofuranosidase, main component was cellulases in these all reactions (Van Dyk and Pletschke 2012). Endo 1,3 beta-D glucanase enzyme coding genes are the only enzyme which characterized endo 1,3,D glucanase found in marine bacteria *Formosa algae* KMM3553 strain and also the only glucanase hydrolases 16 family with transglycosylation activity with the expression in *E coli*. Polysaccharides digests products by the hydrolytic activity of an enzyme which helps to classify and know about the substrate specificity as glucan endo 1,3 beta-D glucosidase enzyme. The amino acid sequences of an enzyme consist of 556 residues with 45–85% of similar sequences with beta 1,3 glucanase of CAZy family 16 glycoside hydrolase from bacterial source. The molecular weight of an enzyme was found to the size of 61 kDa, optimum environmental factors were at pH 5.5, temperature of 45 °C exhibits maximum activity at pH 5.5.45 °C, and half-life period at 45 °C is 20 min, Km for hydrolysis of laminarin, 0.388 mM (Kusaykin et al. 2017).

3.3 Applications

Beta-glucans are the sugars which mainly established in the cell walls of fungi, bacterial, yeasts, algae, lichens and some of the plants including oats and barley, some of these are used as source of medicines. In bacterial cell, metabolic activities of these enzymes were reported in detail for an endo 1,3 beta-glucanase and also 1,3,1,4,beta-glucanases (Fuchs et al. 2003). These endo 1,3 beta-glucanases hydrolyze an internal 1,3 beta glucosyl linkages while some of 1,3,1,4,beta-glucanases can be able to hydrolyze only 1,4 beta-glucanases. The endo 1,3 beta-glucanase from a bacterial origin can show similarities in the sequences of 1,3,1,4,beta-glucanases (Gueguen et al. 1997) and these enzymes exist in glycosyl hydrolase family, i.e., GH16 (Henrissat and Bairoch 1996). 1,3-β-Glucanases are useful for the prevention of the entry of some fungal penetration and protect the plants from these pathogenic agents by their hydrolyzing 1,3-β-glucan (Chen et al. 1993). β-Glucanase plays a main component of many of the cell walls of fungal species and a main storage, structural polysaccharide of macro algae from marine sources. An enzyme glycoside hydrolase β-1,3-glucanase, extensively found in cell wall of bacteria, plants, and fungi, which able to works on 1,3-β-glucosidic bonds of 1,3-β-glucan and transfer or able to hydrolyze glycosides (Kobayashi et al. 2016; Kusaykin et al. 2017) and these beta 1,3 glucanases are capable to hydrolyze or breakdown these beta 1,3 glucans as a main source of defensive agents against fungal pathogens which infect in plants, also can inhibit the growth of fungi in fermentation technology (Jadhav and Gupta 2016; Lee 2012). The beta 1,3 glucans can be capable to degrade into 1-3 linked beta-glucan oligosaccharides having antitumor and immunomodulating activities (Fu et al. 2015). An exo beta 1,3 glucanase enzyme representing gene encoded an immunodominant protein which also acts as a cell wall remodeler, this enzyme has an important applicant for serology diagnostics tests and immune therapeutical application (Ribeiro et al. 2017).

The enzyme 1,3 beta-glucanase is finely constituted for preparing protoplast for cell fusion, transformation and some protein products extraction, etc. like in biotechnology researches and processes (Parrado et al. 1996). In addition, 1,3 beta-glucanase has a growing importance in pharmaceutical purposes by modifying beta-glucan (Wasser 2002). An endo-1,3-β-glucanase purified and characterized from *Arthrobacter* sp. with a molecular weight of 32.5 kDa, also β-glucan oligomer digested by tendo-1,3-β-glucanase from laminarin which has immunologically stimulating activities on human monocytes (Pang et al. 2004) has exclusive characteristics of the hydrolysis of 1,3 beta-glucans with 1,6 β and 1,4 β linkages like lichenan and laminarin also more effective than the other microbial endo 1,3 beta-glucanases (Fuchs et al. 2003; Nobe et al. 2003). Endo 1,3 beta-glucanases from soil bacteria *Paenibacillus* sp. have stronger antifungal effects by damaging the cell wall of phytopathogenic fungi (Hong and Meng 2003). β-Glucanase is used as a cellulase enzyme in some of biological pretreatment method in lignocellulosic ethanol production from cane, used in saccharification of cellulose and ethanol production and has defense mechanism against plant pathogenic fungi.

Beta-glucans are enzymes with various applications, are used mainly as a suitable food additives in some of food items like salads, frozen desserts, sour creams, cheese spreads, etc. Foods with high beta-glucans can always use in low fat diet to prevent some diseases, mainly heart problems. Some researchers reported about the diet of a person who was eating at least around 3.6gm of beta-glucan as a content in their foods like barley/oats or soluble fibers per day is useful to diminish the risk of heart-related diseases. These beta-glucan enzymes are the one form of the dietary fiber and help the patients to decrease the level of bad cholesterol also boosting the health of their heart. FDA (Food and Drug Administration, US) Trusted Source approved some of the foods as heart healthy label, having the presence of large quantities of beta-glucan. Numerous studies reported about the capability of beta-glucans to lower the level of cholesterol and triglycerides in humans. Similar works revealed the results of usage of glucans, persons who consuming at least 3gm beta-glucan per day had reduced the LDL (bad cholesterol) level in their blood at least 5–7%. Some of the researchers believe that the beta-glucan has good effects on the immune system of patients, and more experimental work is necessary to prove the above (Goldman 2016).

An endo 1,3 beta-D glucanase enzyme is the rare enzyme isolated from a marine bacteria *Formosa algae* KMM 3553 and is capable to involve in catalytic reactions of transglycosylation reactions, also this enzyme catalyzed the transfer of glyconic portion of substrate on methyl beta-D xylopyranoside and glycerol, also it may be used for the structural characterization and determination of beta 1,3 glucans or mixed 1,3,14 and 1,3,1,6 beta-glucans and a synthesis of a novel carbohydrate compounds enzymatically (Kusaykin et al. 2017). The purified enzyme mutant E 119G of an enzyme endo 1,3 beta-glucanase having a secondary structure and with foldings shows similarity with the wild type, when comparing with these wild-type enzyme which exhibited more activity on transglycosylation but decreased the hydrolyzing ability. These two activities of E119G, transglycosylation and hydrolytic activities were found to be decreased with decreased range of temperature while the ratio of the transglycosylation products were increased in quantity. Transglycosylation activity was higher in the temperature-dependent degree of reduction and in pH range between 4.0 and 7.4,the activities of these enzymes were similar, while in pH of slightly alkaline, i.e., pH of 8 and 8.5, the enzymatic reaction was slightly slowdown (Hantani et al. 2018). Further studies are necessary to improve transglycosylases activity and specificity and also needed the additional plannings for using these mutant glycosidases which was decreased its hydrolyzing capability and changed into glycosyl transferases, when replacing the catalyzing residues in an enzyme, results show the failure of an hydrolyzing activity of an enzyme. Another reaction of water with an enzyme substrate intermediate, during in the reaction with glucanase, as a result of this reaction hydrolyzed sugar production was take place and the water molecule which was replaced by sugar molecule and glycoside linkage was formed as result of transglycosylation reaction (Hantani et al. 2018).

Cellulosimicrobium cellulans producing an enzyme endo 1,3 beta-glucanase with carbohydrate-binding module and a catalytic domain, the crystalline X ray crystallographic structure of this domain was studied t 1.66 Å resolution and determined the detailed architecture as Beta jelly roll sandwich like folds and found that and has been capable to withstand 6 glucopyranose units (Oda et al. 2018). Endo 1,3 beta-glucanase with a molecular weight of 44KDa was isolated and purified from *Paenibacillus* strain, this enzyme has increased its specific activity with 28-fold with 31% of recovery and catalyzes glucan hydrolysis with 1,3 beta linkage shows endolytic or binding activities to cellulose and xylan like insoluble polysaccharide substrates. Related study was reported on 1,3 beta-glucanase enzyme extracted and purified from an actinomycetes, *Streptomyces sioyaensis*, the antifungal activity was checked, the results showed the inhibition of the growth of fungi, *Pythium aphanidermatum* and *Rhizoctonia solani* are phytopathogens by the cell wall damage due to action of this enzyme. This *Paenibacillus* enzyme had more antifungal effect (Hong et al. 2008).

1,3-1,4-Beta-glucanases hydrolyze linear beta-glucans with beta 1,3 and beta 1,4 linkages with cleavage specificity for beta-1,4 glycosidic bonds. The structure-function aspects of enzymatic action like mechanistic enzymology, protein engineering technologies, and X-ray crystallographic were elucidated (Planas 2001). The glucans with 1,3 beta linkages found in organisms, like bacteria, fungi, plants, algae, etc. were widely distributed in nature. Beta 1,3 linked glucan is major structural polysaccharide (Sova et al. 2013). Related study on laminarinase, also named as beta 1,3 glucanohydrolase/beta 1,3 glucanase, is able to break down and hydrolysis of the enzyme 1,3 and 1,4 linkages in beta-D glucans (Zverlov et al. 2001), also this enzyme can be involved in several functional and physiological processes in some of microbes including bacteria, fungi, etc. and plants (Sun et al. 2000) and also this enzyme has a storage polysaccharide found in macroalgae (Sova et al. 2013; Graiff et al. 2016). In bacterial cells, laminarin enzyme was classified into exo beta 1,3 glucanases and endo beta 1,3 glucanases, and also they are similar sequence pattern with endo beta 1,3 1,4 glucanases. Glucose with its beta 1,3 linkages were connected by their partial branches with the beta 1,6 linkages in the main chain and reported that this enzyme shows biological activities like antiapoptotic and antitumor activities (Kim et al. 2006; Kuda et al. 2005). Laminarin plays a major role in the fermentation production of bioethanol, due to the presence of sugars which can be able to involve in fermentation process (Harris et al. 2014) and also is useful in design of biocatalysts for developing process for the conversion of laminarin to fermentative sugars (Takeda et al. 2011; Radakovits et al. 2010). Some extracellular polysaccharides like curdlan and lichenin are nontoxic sugar compounds, which are used in many food and pharmaceutical industries, etc. (Zhan et al. 2012).

Curdlan, the unbranched polysaccharide, its polymerization is about 135 glucose residues, which may consisting with a glucosyl residues are linked with beta-D 1,3 bonding (Harada et al. 1968). Lichenin is sharing a symbiotic relationship, which has a ratio of 2: 3: 1 with beta-D 1,4 glucopyranosyl to the beta-D 1,4 glucopyranosyl residues found in *Cetraria islandica* strain (Honegger 2001). Exo beta 1,3 glucanase/endo beta 1,3 glucanase is capable to hydrolyze laminarin like beta 1,3 glucanase and the reaction results in the release of glucose otherwise oligosaccharide residues while endo-β-1,3-1,4-glucanases may also hydrolyze polysaccharides (Qin et al. 2017). Beta 1, 3,1,4 glucanases are the enzymes becoming a potent biological catalysis and may contributing more effective processes in the time of biochemicals and bioenergy (Wargacki et al. 2012; Planas 2000).

3D structure and the enzyme specificity were determined with the help of the Carbohydrate Active Enzymes Database, CAZy, in which the sequence-based classification of enzymes with family linking of each enzyme was provided (Lombard et al. 2014), some members of laminarinases from bacterial sources belong in the family of GH16. These enzymes were well characterized and also analyzed the crystalline structure of an enzyme isolated from *Thermotoga maritima* (Jeng et al. 2011), hyperthermophile *Pyrococcus furiosus* (Ilari et al. 2009; Van Lieshout et al. 2004), *Nocardiopsis* sp. F9623 strain, and *Zobellia galactanivorans* (Labourel et al. 2013). It is difficult for classifying the family of GH16 enzymes from beta jelly roll folds with two anti-parallel beta sheets (Stam et al. 2006; Zhan et al. 2012). *Flavobacterium* sp. UMI-01 strain is a newly identified novel bacterium which was isolated from a decayed brown pigmented algae growing in the nutrient medium with alginate or laminarin identified from decayed brown algae, grown in medium containing alginate/laminarin as a source of carbon (Inoue et al. 2014). The bacterium contains the higher ratio of GH16 laminarinases family and is environmentally important in the decomposition of the biomass of algae. Genomic analysis of enzyme sequences disclosed the existence of the gene codes for GH16 family of beta 1,3 laminarinase and which is named as ULam111. Two hundred and thirty-five residues constitute this enzyme and have the molecular weight of around 27KDa. The structural analysis of an enzyme revealed the capability on thermostability and substrate recognition ability mainly for specific polysaccharides and their degradations. These enzymes have many applications in various areas like brewing of beer, food and feed additives, etc. (Qin et al. 2017). *Fusobacterium nucleatum* and *Tannerella forsythia* are the causative bacterial agents for dental plaques, they have more genetic loci for TonB-dependent membrane receptors, TDR for the acquisition for glycans. The BFO 0186–BFO 0188 is a locus, which encodes the transporter TDR-SusD with the potential candidate beta-glucan and its hydrolyzing enzyme BFO_0186 expresses β-glucanase which cleaves glucans with β-1,6 and β-1,3 linkages. This loci is with the induction of an enzyme activity of beta-glucanase in co-biofilms which formed from *T. forsythia* and *F. nucleatum.* In some of the mixed colonies of biofilms, beta-glucanase activity which enhanced the hydrolytic reaction of beta-glucan and formed oligomers and monomers as a result of reaction, which were used as nutrients by *F. nucleatum* (Honma et al. 2017).

The amino acid sequences of endo beta 1,3 glucanase of alkaliphilic strain of *Nocardiopsis* sp. were analyzed, results show the highest similarities to GHF16 endo beta 1,3 glucanase family, also this enzyme exhibits eight-fold more activity towards beta 1,3,1,4 glucanases when compared to beta 1,3 glucan, its optimum activity was at 70^0 C and at pH 6 (Masuda et al. 2006). These results revealed that, the beta-glucanase Family would enzyme endo 1,3 beta-glucanase (Masuda et al. 2006). Researchers were structurally analyzed the crystalline composition of an enzyme from wild-type *Bacillus* strains, *Bacillus licheniformis* (Hahn et al. 1995a, b, c) and *B. macerans* (Hahn et al. 1995a, b, c); this enzyme is an endo beta 1,3,1,4 glucanases (Hahn et al. 1994, 1995a, b, c; Aÿ et al. 1998a, b) and one of the crystal structure of endo beta 1,3,1,4 beta-glucanase enzyme, which was circularly permuted enzyme isolated from *Fibrobacter succinogenes* (Tsai et al. 2003a, b) was analyzed and results found to be these enzymatic structure of bacterial enzyme has not yet reported.

Various researches reported on the usage of beta-glucans, which have an ability to reduce the higher risk of infections like sepsis in trauma patients and also some of the beta-glucans produced by yeast were used as injection through vein to patients after surgery to lower the chances of infection. Some of the studies related to beta glucan in Hay Fever that, taking beta-glucan for almost 4 weeks to decrease the symptoms in patients. There is some evidence that giving a specific kind of beta-glucans as an injection into the vein can extend life in women with advanced cervical cancer when used with standard cancer treatment.

References

Asano T, Taki J, Yamamoto M, Aono R (2014) Cloning and structural analysis of bglm gene coding for the fungal cell wall-lytic β-1,3-glucan-hydrolase bglm of *Bacillus circulans* IAM 1165. Biosci Biotechnol Biochem 66:1246–1255

Aÿ J, Götz F, Borriss R, Heinemann U (1998a) Structure and function of the *bacillus* hybrid enzyme GluXyn-1: native-like jellyroll fold preserved after insertion of autonomous globular domain. Proc Natl Acad Sci U S A 95:6613–6618

Aÿ J, Hahn M, Decanniere K, Piotukh K, Borriss R, Heinemann U (1998b) Crystal structures and properties of de novo circularly permuted 1,3-1,4-β-glucanases. Proteins 30:155–167

Barras DR, Stone BA (1969) Beta-1,3-glucan hydrolases from *Euglena gracilis*. The nature of the hydrolases. Biochim Biophys Acta 191(2):329–341

Beckmann L, Simon O, Vahjen W (2006) Isolation and identification of mixed linked beta -glucan degrading bacteria in the intestine of broiler chickens and partial characterization of respective 1,3-1,4-beta -glucanase activities. J Basic Microbiol 46(3):175–185

Beckmann L, Vahjen W, Simon O (2000) Isolation of an 1,3-1,4-β-glucan degrading *enterococcus faecium* strain from the intestinal tract of chicken and partial characterization of its novel 1,3-1,4-β-glucanase. J Basic Microbiol 40(5-6):303–310. https://doi.org/10.1002/1521-4028(200012)40:5/6<303

Cecchini DA, Pepe O, Pennacchio A, Fagnano M, Faraco V (2018) Directed evolution of the bacterial endo-β-1,4-glucanase from *Streptomyces* sp. G12 towards improved catalysts for lignocellulose conversion. AMB Express 8(1):1–10

Chen L, Fincher GB, Hùj PB (1993) Evolution of polysaccharide hydrolase substrate specificity. Catalytic amino acids are conserved in barley 1, 3-1, 4-and 1, 3-beta-glucanases. J Biol Chem 268:13318–13326

Cheng YM, Hong TY, Liu CC, Meng M (2009) Cloning and functional characterization of a complex endo-beta-1,3-glucanase from *Paenibacillus* sp. Appl Microbiol Biotechnol 81(6): 1051–1061. https://doi.org/10.1007/s00253-008-1617-9. Epub 2008 Sep 19.PMID: 18802694

Déjean G, Tamura K, Cabrera A, Jain N, Pudlo NA, Pereira G, Viborg AH, Petegem FV, Martens EC, Brumer H (2020) Synergy between cell surface Glycosidases and glycan-binding proteins dictates the utilization of specific Beta(1,3)- Glucans by human gut Bacteroides. MBio 11(2): e00095–e00020. https://doi.org/10.1128/mBio.00095-20

Dodson G, Wlodawer A (1998) Catalytic triads and their relatives. Trends Biochem Sci 23:347–352

Doolittle RF (1994) Convergent evolution: the need to be explicit. Trends Biochem Sci 19:15–18

Fu Y, Cheng L, Meng Y, Li S, Zhao X, Du Y, Yin H (2015) *Cellulosimicrobium cellulans* strain E4-5 enzymatic hydrolysis of curdlan for production of (1→3)-linked β-D-glucan oligosaccharides. Carbohydr Polym 134:740–744

Fuchs KP, Zverlov VV, Velikodvorskaya GA, Lottspeich F, Schwarz WH (2003) Microbiology 149:1021–1031

Gaiser OJ, Piotukh K, Ponnuswamy MN, Planas A, Borriss R, Heinemann U (2006) Structural basis for the substrate specificity of a *bacillus* 1,3-1,4-beta-glucanase. J Mol Biol 357(4):1211–1225. https://doi.org/10.1016/j.jmb.2006.01.014

Rena Goldman (2016) Beta Glucan: the heart-healthy fiber. Healthline. Medically reviewed by Natalie Butler, R.D., L.D

Graiff A, Ruth W, Kragl U KU (2016) Chemical characterization and quantification of the brown algal storage compound laminarin–a new methodological approach. J Appl Phycol 28:533–543

Gueguen Y, Voorhorst WGB, van der Oost J, de Vos WM (1997) J Biol Chem 272:31258–31264

Hahn M, Keitel T, Heinemann U (1995c) Crystal and molecular structure at 0.16-nm resolution of the hybrid *bacillus* endo-1,3-1,4-β-D-glucan 4-glucanohydrolase H(A16-M). Eur J Biochem 232:849–858

Hahn M, Olsen O, Politz O, Borriss R, Heinemann U (1995a) Crystal structure and site-directed mutagenesis, Endo1,3,-1,4,- beata glucanase. J Biol Chem 270(7):3081–3088. https://doi.org/10.1074/jbc.270.7.3081

Hahn M, Piotukh K, Borriss R, Heinemann U (1994) Native-like *in vivo* folding of a circularly permuted jellyroll protein shown by crystal structure analysis. Proc Natl Acad Sci U S A 91: 10417–10421

Hahn M, Pons J, Planas A, Querol E, Heinemann U (1995b) Crystal structure of *bacillus licheniformis* 1,3-1,4-β-D-glucan 4-glucanohydrolase at 1.8 Å resolution. FEBS Lett 374: 221–224

Håkansson K, Wang AH-J, Miller CG (2000) The structure of aspartyl dipeptidase reveals a unique fold with a Ser-his-Glu catalytic triad. Proc Natl Acad Sci U S A 97:14097–14102

Hantani Y, Motoki S, Miyagawa A, Yamamura H, Oda M (2018) Transglycosylation activity of catalytic domain mutant of Endo-1,3-β-glucanase from *Cellulosimicrobium cellulans*. Protein Pept Lett 25(8):734–739. https://doi.org/10.2174/0929866525666180704094109

Harada T, Misaki A, Saito H (1968) Curdlan: a bacterial gel-forming β-1,3-Glucan. Arch Biochem Biophys 124:292–298

Harris PV, Xu F, Kreel NE, Kang C, Fukuyama S (2014) New enzyme insights drive advances in commercial ethanol production. Curr Opin Chem Biol 19:162–170

Henrissat B, Bairoch A (1996) Biochem J 316:695–696

Honegger R, Haisch (2001) A Immunocytochemical location of the (1,3, 1,4)-β-glucan lichenin in the lichen-forming ascomycete *Cetraria islandica* (Icelandic moss). New Phytol 150:739–746

Hong TY, Cheng CW, Huang JW, Meng M (2002) Isolation and biochemical characterization of an endo-1,3-β-glucanase from *Streptomyces sioyaensis* containing a C-terminal family 6 carbohydrate-binding module that binds to β-1,3-glucan. Microbiology 148:1151–1159

Hong TY, Hsiao YY, Meng M, Li TT (2008) The 1.5 a structure of endo-1,3-beta-glucanase from *Streptomyces sioyaensis:* evolution of the active-site structure for 1,3-beta-glucan-binding specificity and hydrolysis. Acta Crystallogr D Biol Crystallogr 9:964–970. https://doi.org/10.1107/S0907444908021550

Hong TY, Meng M (2003) Biochemical characterization and antifungal activity of an endo-1,3-β-glucanase of *Paenibacillus* sp. isolated from garden soil. Appl Microbiol Biotechnol 61: 472–478

Honma K, Ruscitto A, Sharma A (2017) β-Glucanase activity of the Oral bacterium *Tannerella forsythia* contributes to the growth of a partner species, *Fusobacterium nucleatum*, in Cobiofilms. Appl Environ Microbiol 84(1):e01759–e01717. https://doi.org/10.1128/AEM.01759-17

Ilari A et al (2009) Crystal structure of a family 16 endoglucanase from the hyperthermophile *Pyrococcus furiosus*-structural basis of substrate recognition. FEBS J 276:1048–1058

Inoue A et al (2014) Characterization of an alginate lyase, FlAlyA, from *Flavobacterium* sp. strain UMI-01 and its expression in *Escherichia coli.* Mar Drugs 12:4693–4712

Jadhav SB, Gupta A (2016) Studies on application of β-1,3 glucanase in the degradation of glucans produced by *Botrytis cinerea* and inhibition of fungal growth. Biocatal Agric Biotechnol 7:45–47

Javmen A, Grigiskis S, Rudenkov M, Mauricas M (2013) Purification and partial characterization of a novel β-1, 3-endoglucanase from *Streptomyces rutgersensis*. Protein J 32(5):411–417. https://doi.org/10.1007/s10930-013-9500-7

Jeng WY, Wang NC, Lin CT, Shyur LF, Wang AHJ (2011) Crystal structures of the laminarinase catalytic domain from *Thermotoga maritima* MSB8 in complex with inhibitors. J Biol Chem 286:45030–45040

Jens W, Mangold M, Tannock GW (2005) Construction, analysis, and β-Glucanase screening of a bacterial artificial chromosome library from the large-bowel microbiota of mice. Evolutionary and Genomic Microbiology 71(5):2347–2354. https://doi.org/10.1128/AEM.71.5.2347-2354

Kim KH, Kim YW, Kim HB, Lee BJ, Lee DS (2006) Anti-apoptotic activity of laminarin polysaccharides and their enzymatically hydrolyzed oligosaccharides from Laminaria japon-Ica. Biotechnol Lett 28:439–446

Kobayashi T, Uchimura K, Kubota T, Nunoura T, Deguchi S (2016) Biochemical and genetic characterization of β-1,3 glucanase from a deep subsea floor *Laceyella putida*. Appl Microbiol Biotechnol 100:203–214

Kuda T, Yano T, Matsuda N, Nishizawa M (2005) Inhibitory effects of laminaran and low molecular alginate against the putrefactive compounds produced by intestinal microflora *in vitro* and in rats. Food Chem 91:745–749

Kusaykin MI, Belik AA, Kovalchuk SN, Dmitrenok PS, Rasskazov VA, Isakov VV, Zvyagintseva TN (2017) A new recombinant endo-1,3-β-D-glucanase from the marine bacterium *Formosa algae* KMM 3553: enzyme characteristics and transglycosylation products analysis. World J Microbiol Biotechnol 33:40

Labourel A et al (2013) The β-glucanase ZgLamA from *Zobellia galactanivorans* evolved a bent active site adapted for efficient degradation of algal laminarin. J Biol Chem 289:2027–2042

Lee SY (2012) Biocontrol of anthracnose in pepper using chitinase, β-1,3 glucanase, and 2-furancarboxaldehyde produced by *Streptomyces cavourensis* SY224. J Microbiol Biotechnol 22:1359–1366

Lee KC, Arai T, Ibrahim D, Kosugi A, Prawitwong P, Lan D, Murata Y, Mori Y (2014) Purification and characterization of a thermostable Laminarinase from *Penicillium rolfsii* c3-2(1) IBRL. Bio Resources 9(1):1072–1084

Lombard V, Golaconda Ramulu H, Drula E, Coutinho PM, Henrissat B (2014) The carbohydrate-active enzymes database (CAZy) in 2013. Nucleic Acids Res 42:D490–D495

Malet C, Planas A (1997) Mechanism of *bacillus* 1,3-1,4-beta-D-glucan 4-glucanohydrolases: kinetics and pH studies with 4-methylumbelliferyl beta-D-glucan oligosaccharides. Biochemistry 36(45):13838–13848

Masuda S, Endo K, Koizumi N, Hayami T, Fukazawa T, Yatsunami R, Fukui T, Nakamura S (2006) Molecular identification of a novel β-1,3-glucanase from alkaliphilic *Nocardiopsis* sp. strain F96. Extremophiles 10:251–255

Miyanishi N, Hamada N, Kobayashi T, Imada C, Watanabe E (2003) Purification and characterization of a novel extracellular β-1,3-glucanase produced by *Bacillus clausii* NM-1 isolated from ezo abalon *Haliotis discus hannai*. J Biosci Bioeng 95:45–51

Nobe R, Sakakibara Y, Fukuda N, Yoshid N, Ogawa K, Suiko M (2003) Biosci Biotechnol Biochem 67:1349±1357

Oda M, Inaba S, Kamiya N, Bekker G-J, Mikami B (2018) Structural and thermodynamic characterization of endo-1,3-β-glucanase: insights into the substrate recognition mechanism. Biochim Biophys Acta Proteins Proteom 1866(3):415–425. https://doi.org/10.1016/j.bbapap.2017.12.004

Okazaki K, Nishimura N, Matsuoka F, Hayakawa S (2007) Cloning and characterization of the gene encoding endo-β-1,3-glucanase from *Arthrobacter* sp. NHB-10. Biosci Biotechnol Biochem 71:1568–1571

Pang Z, Kang YN, Ban M, Oda M, Kobayashi R, Ohnishi M, Mikami B (2005) Crystallization and preliminary crystallographic analysis of endo-1,3-beta-glucanase from *Arthrobacter* sp. Acta Cryst 61:68–70

Pang Z, Otaka K, Suzuki Y, Goto K, Ohnishi M (2004) J Biol Macromol 4:57–66

Parrado J, Escuredo PR, Conejero-Lara F, Kotik M, Ponting CP, Asenjo JA, Dobson CM (1996) Biochim Biophys Acta 1296:145–151

Pérez-Avalos O, Sánchez-Herrera LM, Salgado LM, Ponce-Noyola T (2008) A bifunctional endoglucanase/endoxylanase from *Cellulomonas flavigena* with potential use in industrial processes at different pH. Curr Microbiol 57(1):39–44. https://doi.org/10.1007/s00284-008-9149-1

Planas A (2000) Bacterial 1,3-1,4-beta-glucanases: structure, function and protein engineering. Biochim Biophys Acta 1543(2):361–382. https://doi.org/10.1016/s0167-4838(00)00231-4

Qin H-M, Miyakawa T, Inoue A, Nakamura A, Nishiyama R, Ojima T, Tanokura M (2017) Laminarinase from *Flavobacterium* sp. reveals the structural basis of thermostability and substrate specificity. Sci Rep 7(1):1–9

Radakovits R, Jinkerson RE, Darzins A, Posewitz MC (2010) Genetic engineering of algae for enhanced biofuel production. Eukaryot Cell 9:486–501

Ribeiro TC, Weiblen C, de Azevedo MI, de Avila BS, Robe LJ, Pereira DI, Monteiro DU, Lorensetti DM, Santurio JM (2017) Microevolutionary analyses of *Pythium insidiosum* isolates of Brazil and Thailand based on exo-1,3-β-glucanase gene. Infect Genet Evol 48:58–63

Shi P, Yao G, Yang P, Li N, Luo H, Bai Y, Wang Y, Yao B (2010) Cloning, characterization, and antifungal activity of an endo-1,3-beta-D: -glucanase from *Streptomyces* sp. S27. Appl Microbiol Biotechnol 85(5):1483–1490. https://doi.org/10.1007/s00253-009-2187-1

Singh RP, Rajarammohan S, Thakur R, Hassan M (2020) Linear and branched *β*- Glucans degrading enzymes from versatile *Bacteroides uniformis* JCM 13288 T and their roles in cooperation with gut bacteria. Gut Microbes 12(1):1–18. https://doi.org/10.1080/19490976.1826761

Sova VV, Pesentseva MS, Zakharenko AM, Kovalchuk SN, Zvyagintseva TN (2013) Glycosidases of marine organisms. Biochemistry Moscow 78:746–759

Stam MR, Danchin EG, Rancurel C, Coutinho PM, Henrissat B (2006) Dividing the large glycoside hydrolase family 13 into subfamilies: towards improved functional annotations of alpha-amylase-related proteins. Protein Eng Des Sel 19:555–562

Sun L, Gurnon JR, Adams BJ, Graves MV, Etten JLV (2000) Characterization of a β-1,3-glucanase encoded by chlorella virus PBCV 1. Virology 276:27 36

Takeda H, Yoneyama F, Kawai S, Hashimoto W, Murata K (2011) Bioethanol production from marine biomass alginate by metabolically engineered bacteria. Energy Environ Sci 4:2575–2581

Tanabe Y, Oda M (2011) Molecular characterization of endo-1,3-β-glucanase from *Cellulosimicrobium cellulans*: effects of carbohydrate-binding module on enzymatic function and stability. BBA 1814:1713–1719

Tsai LC, Amiraslanov I, Chen HR, Chen YW, Lee HL, Liang PH, Liaw YC (2015) Structures of exoglucanase from *clostridium cellulovorans*: cellotetraose binding and cleavage. Acta Crystallogr F Struct Biol Commun 71(10):1264–1272. https://doi.org/10.1107/S2053230X15015915

Tsai L-C, Shyur L-F, Lee S-H, Lin S-S, Yuan HS (2003a) Crystal structure of a natural circularly permuted jellyroll protein: 1,3-1,4-beta-D-glucanase from *Fibrobacter succinogenes*. J Mol Biol 330(3):607–620. https://doi.org/10.1016/s0022-2836(03)00630-2

Tsai LC, Shyur LF, Lee SH, Lin SS, Yuan HS (2003b) Crystal structure of a natural circularly permuted jellyroll protein: 1,3-1,4-β-D-glucanase from *Fibrobacter succinogenes*. J Mol Biol 330:607–620

Van Dyk JS, Pletschke BI (2012) A review of lignocellulose bioconversion using enzymatic hydrolysis and synergistic cooperation between enzymes-factors affecting enzymes, conversion and synergy. Biotechnol Adv 30(6):1458–1480. https://doi.org/10.1016/j.biotechadv.2012.03.002

Van Lieshout J, Faijes M, Nieto J, Van Der Oost J, Planas A (2004) Hydrolase and glycosynthase activity of endo-1,3-β-glucanase from the thermophile *Pyrococcus furiosus*. Archaea 1(4):285–292

Vincken JP, Beldman G, Voragen AG (1997) Substrate specificity of endoglucanases: what determines xyloglucanase activity? Carbohydr Res 298(4):299–310

Wargacki AJ et al (2012) An engineered microbial platform for direct biofuel production from brown macroalgae. Science 335:308–313

Wasser S (2002) Appl Microbiol Biotechnol 60:258–274

Wu Q, Dou X, Wang Q, Guan Z, Cai Y, Liao X (2018) Isolation of β-1,3-Glucanase-producing microorganisms from *Poria cocos* cultivation soil via molecular biology. Molecules 23(7):1555. https://doi.org/10.3390/molecules23071555

Zhan XB, Lin CC, Zhang HT (2012) Recent advances in curdlan biosynthesis, biotechnological production, and applications. Appl Microbiol Biotechnol 93:525–553

Zhou J, Li Z, Wu J, Li L, Li D, Ye X, Luo X, Huang Y, Cui Z, Cao H (2017) Functional analysis of a novel β-(1,3)–Glucanase from *Corallococcus* sp. strain EGB containing a Fascin-like module. Appl Environ Microbiol 83(16):e01016–e01017. https://doi.org/10.1128/AEM.01016-17

Zverlov VV, Volkov IY, Velikodvorskaya GA, Schwarz WH (2001) The binding pattern of two carbohydrate-binding modules of laminarinase Lam16A from *Thermotoga neapolitana*: differences in β-glucan binding within family CBM4. Microbiology 147:621–629

Chapter 4
Beta-Glucanases: Sources and Production from Fungi

Shweta Kulshreshtha and Vinay Sharma

Abstract β-glucanases are the key enzymes involved in the conversion of complex polysaccharides such as glucan, cellulose into simple units of glucose by the hydrolysis of the glycosidic bond. Glucanases are reported to be having multifarious applications like degradation of complex polymers into simple compounds, flavour enhancer of wines and composting of agricultural wastes. Therefore, the importance of glucanases cannot be ignored. Recent research has focused on the production of β-glucanases using low-cost materials and different microbial strains. Many microbial strains have been reported to produce a high amount of glucanases. This chapter focuses on the different forms of fungi to produce glucanases such as fungi, yeast and mushroom forms. This chapter also describes the different mushrooms as a possible source of production. The functions and production of alpha- and β-1,3-glucanases are discussed in the detail in this review. We also shed light on different applications of glucanases considering traditional and current knowledge.

4.1 Introduction

The growing demand for a variety of enzymes is bringing out major research in the field of enzyme technology. Many biotechnological approaches are based on the application of enzymes including the production of food and food products, feedstock for animals, development in cosmetics, a tool for the fermentation industry and research and development of new era products. Different microbes are rich in various enzymes and therefore act as the source of enzymes. Again, biotechnological approaches play a major role in isolation, purification, characterization and further, their use in different applications.

Like the other enzymes, glucanases play an important role in many industrial processes such as food and food products-based industries, biocontrol agent production-based industries, diagnostics, therapeutics, etc. due to their ability to

S. Kulshreshtha · V. Sharma (✉)
Amity Institute of Biotechnology, Amity University Rajasthan, Jaipur, India

N. S. Pradeep, L. K. Edison (eds.), *Microbial Beta Glucanases*, Interdisciplinary Biotechnological Advances, https://doi.org/10.1007/978-981-19-6466-4_4

cleave glucans. Glucan is a complex polymer made up of glucose residues linked by β-glycosidic linkages. Glucanases play a pivotal role in the cleavage of non-cellulolytic β-glucans. These are produced by several organisms like bacteria, fungi, algae, molluscs and higher plants. This chapter shed light on the sources, production and applications of different fungal glucanases. Their role in many processes is also discussed in light of knowledge present in literature.

4.2 Types of Glucanases

The glucanases are majorly two types, i.e. alpha-glucanase and β-glucanase. β-glucanases are present in different fungi including yeasts and mushrooms with different hydrolytic activity, i.e. exo-lytic or endolytic mechanisms. β-1,3-glucanases and β-1,6-glucanases and many more secreted by different fungi are mentioned in Fig. 4.1.

Exo-β-1,3-glucanases (EC 3.2.1.58) hydrolyse the substrate by sequentially cleaving glucose residues from non-reducing ends.

Endo-β-1,3-glucanases (EC 3.2.1.39) cleave β linkages at random sites along the polysaccharide chain and release short oligosaccharides chains.

B-D-1,6-glucanases cleave β-1,6-glucan and make it available for plant nutrition.

Endo-(1→4)-β-glucanases (cellulases) cleave cellulose randomly throughout the cellulosic chain.

β-1,3-1,4-glucanases (or lichenases; EC 3.2.1.73) hydrolyse both β-1,3 and β-1,4-glucosidic linkages for diverse applications in food, feed and detergent industries.

Alpha-1,3-glucanases cleave alpha-1,3-glucan, require as endogenous carbon source and virulence factor. Alpha-1,3-glucans are present in the fungi as a constituent of the cell wall.

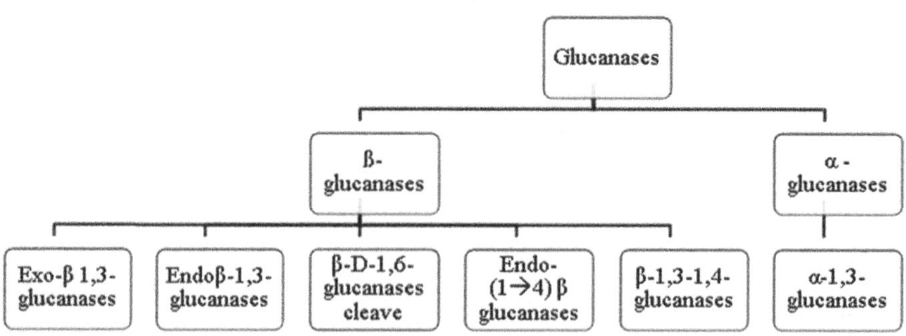

Fig. 4.1 Different types of glucanases

4.3 Glucanases in Multicellular Fungi

Fungi are known to produce a wide range of glucanases which may be extracellular or intracellular, endo- or exo-type. These can be produced by a single fungus and may share a common linkage to attack. Generally, genes are responsible for the genetic diversity of enzymes. However, the diversity of glucanases is the result of a post-translational modification. The production of multiple β-glucanase is possible in fungi for β-1,3 glucanases nevertheless, the production of multiple forms of β-1,6-glucanase is not possible by a single fungus (Martin et al. 2007).

4.3.1 Role of Fungal Glucanases

4.3.1.1 Fungal β-1,3-Glucanases

Endo-β-1,3-glucanases (EC 3.2.1.39) cleave β linkages at random sites along the polysaccharide chain and releases short oligosaccharides chains. A different type of endoglucanase, i.e. E*ndo-β-1,3-glucanases (EC 3.2.1.6)*, is reported in *Mucor hiemalis* but have had the same mode of action (Miyazaki et al. 1977).

4.3.1.1.1 Provide Nutrition

Fungal β-1,3-glucanases play an important role in *proving nutrition* to the growing fungal mycelium (Noronha and Ulhoa 2000). These glucanases degrade the glucans and release glucose units which act as a carbon source for the growing mycelium. Thus, it provides nutrition to the growing cells.

4.3.1.1.2 Morphogenesis and Conidial Germination

Fungi occur in two forms: multicellular mycelial form and unicellular yeast form. Their morphological transformation from yeast to mycelial form or conidial germination depends upon the softening of the cell wall (β-1,3-glucan). The cell wall glucan is dissolved due to the action of β-1,3-glucanases (Adams 2004). Therefore, β-1,3-glucanases play an important role in conidial germination and morphogenesis. The morphogenesis of *Aspergillus fumigatus* and the role of glucanases were reported by Mouyna et al. (2013). The glucanases are also important for the organization of proper cell walls as well as for the segregation of conidia. This process is controlled by ENG1 and ENG2–5 genes which code for glucanases. The mutant for this gene showed non-segregation of conidia and improper cell wall development revealing the importance of glucanases in normal conidial development (Mouyna et al. (2013). According to Millet et al. (2019), the genes of GH55 is responsible for conidial germination and normal spreading of conidia.

4.3.1.1.3 As Biocontrol Agent

β-1,3-glucanases, produced by *Trichoderma* species, are active against many fungi and due to their antagonistic action, these act as a biocontrol agent in controlling the fungal infection of the plants (Noronha and Ulhoa 2000). An endo 13-1,3-glucanase was isolated from fungus *Rhizoctonia solani*; and mushroom *Lentinus edodes* and *Agaricus bisporus* which was considered as thaumatin-like protein based on sequence homology. This endo 13-1,3-glucanase is bound strongly to polymeric 13-1,3-glucans and hydrolyse it. Thaumatin-like protein derived from *Rhizoctonia* can degrade *Saccharomyces* after binding with its polymeric glucans (Grenier et al. 2000). Therefore, β-1,3-glucanases can be used for developing biocontrol agents.

4.3.1.1.4 Establishing Infection

β-1,3-glucanases also help in establishing infection during mycoparasitism. Mycoparasitism is the phenomenon of causing infection in one fungus (host) by the other fungi (parasite). The cell wall of fungi is made up of β-1,3-glucans which must be degraded for the entry of other parasitic fungi. To break down the cell wall and get entry into the fungal cell, β-1,3-glucanases are produced to lyse the cell wall. This is required for the penetration of pathogenic fungi into host fungi during mycoparasitism (Inbar and Chet 1997). *Trichoderma harzianum* CECT 2413 is a mycoparasitic fungus that establishes infection by using glucanases. This fungus secretes extracellular β-1,3-glucanases (mol. Wt. 78 kDa) for cleaving β-1,3-glucan in endolytic fashion (Cruz et al. 1995).

4.3.1.1.5 Establishing Endophytes

The aforesaid mentioned process is also required for the establishment of infection by endophytes. Moreover, this mechanism is also required to develop transgenic plants. In a study, conducted by Prasetyawan et al. (2017), transgenic citrus plants were developed by transferring *Trichoderma* L-1 which is endophytic fungi with both glucanases and chitinase activity. The expression of glucanases and chitinase was assessed in artificial media and it was reported that the addition of laminarin and cell wall of phytophthora increases glucanases activity. However, chitin addition increases chitinase activity (Prasetyawan et al. 2017).

4.3.1.1.6 Elicitors of Plant Defence Mechanism

When a plant acquires fungal infection, β-1,3-glucanases are produced at the infection site in plants which act as an elicitor of defence mechanism. The concentration of β-1,3-glucanases was reported to be higher at the infection site in plants. These β-1,3-glucanases lysed the cell wall of pathogenic fungi and released short fragments which are elicitors of plant defence mechanism. EgGlc1-1 gene codes for glucanases and their expression. The expression of this gene was increased by *Trichoderma harzianum* and suppressed by *Ganoderma boninense* in the roots of oil palm seedlings. This revealed the induction of plant defence against *Trichoderma* infection but low activity against *Ganoderma* (Yeoh et al. 2012).

4.3.1.2 β-D-1,6-Glucanases

β-D-1,6-glucanases act on β-D-1,6-glucan and cleave it in a random manner which generates β-D-1,6-linked oligosaccharides. Although β-D-1,6-glucan occurs rarely in nature, β-D-1,6-glucanases are produced by several fungi for the lysis of β-D-1,6-glucans. It is produced with the other glucanases nevertheless, separated from other β-D-glucanases by zone electrophoresis (Reese et al. 1962). An endoglucanase endo-(1→ 6)-β-D-glucanase (EC 3.2.1) was isolated from *Mucor hiemalis* (Miyazaki and Oikawa 1976).

4.3.1.2.1 Role in Establishing Endophytes

Neotyphodium sp. is a fungal endophyte of *Poa ampla* which secretes β-1,6-glucanase in apoplast of plant roots. It showed sequence homology to the enzymes produced by *Trichoderma harzianum* and *Trichoderma virens*. This glucanase helps the fungi in acquiring nutrition by dissolving the glucan content. The presence of β-1,6-glucan can induce the production of glucanases for their breakdown. The β-1,6-glucanase was detected not only in *Poa ampla* but also in several types of grass infected with *Neotyphodium* sp. (Moy et al. 2002).

4.3.1.2.2 Role in Disease Prevention

Verticillium fungicola is a mycoparasitic fungus of *Agaricus bisporus* which produces β-I,6-glucanases. The production of glucanases in artificial media requires the supplementation of chitin or host cell walls. Genetically, the production of β 1,6-glucanase in *V. fungicola* is controlled by the action of *VfGlu1* gene. Agrobacterium-mediated transfer of genes in *V. fungicola* disrupted the normal function of *VfGlu1* gene due to the sharing of the same locus. This leads to the mutation in *VfGlu1* gene which further affects glucanases production. This makes

V. fungicola incapable of infecting *Agaricus bisporus*. This leads to the prevention of infection of *Agaricus* by *V. fungicola* (Amey et al. 2003).

4.3.1.3 Endo-(1→4)-β-Glucanases

This type of endoglucanase was reported into the culture filtrate of *Aspergillus japonicus*. It was reported to be very similar in its activity to bacterial lichenase of *Bacillus subtilis*. Unlike bacterial glucanase, it possesses β-(1,3)-glucoside linkage at a non-reducing end. Sequence analysis revealed its sequence identity to *Aspergillus kawachii* rather than *Aspergillus niger* (Grishutin et al. 2005).

4.3.1.4 Alpha-1,3-Glucanase

As mentioned earlier, these glucanases act on alpha-1,3-glucans and cleave it for a variety of purposes that include morphogenesis during sexual development and virulence factor for establishing infection. Alpha-1,3-glucans are water-insoluble polymers of glucose. It is a homopolymer of glucose linked by alpha1-1,3-linkages. It is present in many fungi such as *Penicillium aurantiogriseum, fusarium oxysporum, Rhizoctonia solani* and *cryptococcus neoformans* (Suyotha et al. 2016).

4.3.1.4.1 Role in Growth and Development

The presence of alpha-glucanases in fungi reveals its specific role. For example, it cleaves alpha-glucan into glucose units which is available as a carbon source for growth and development. This process also occurs during the sexual development of fungi (Zonneveld 1972).

4.3.1.4.2 Role in Establishing Infection

Glucanases play an important role in establishing or preventing infection. The plant produces many enzymes in response to the infection caused by fungi and degrades fungal cells. However, fungi use their glucan content as a defence material against the plant and shield its cell wall by the deposition of alpha-glucans. The plant produces lytic enzymes in response to infection but is unable to cleave alpha-glucans deposited on the fungal cell wall. Hence, fungi survive inside the plant cell and cause infection. The role of alpha-glucans as a virulence factor is established for *M. oryzae* (Fujikawa et al. 2012).

4.3.2 *Production of Fungal Glucanases*

Fungal glucanases are produced at an industrial scale for their commercial use. To increase the production hit and trial method is adopted for optimization of conditions. The conditions can also be optimized using full factorial design or Taguchi method. The steps involved in glucanases production from fungi are as follows:

4.3.2.1 Selection of Suitable Organism

The selection of organisms depends on their ability to produce glucanases. High yield varieties with low production time can be screened for production by fermentation technology. When the specific microbial inoculum is added to a standardized sterile liquid medium for fermentation, glucanases are produced by the microbes under optimal growth conditions. Optimum growth conditions can be standardized by full factorial design method, multivariate statistical approach response surface methodology, etc. for various parameters like pH, temperature, supply of carbon dioxide and different nutrients.

4.3.2.2 Selection of Suitable Substrate

The selection of substrate poses a great influence on the production of glucanases. For instance, the production of β-1,3-glucanases was induced by the supplementation of fungal cell wall polymer or autoclaved dead yeast cells. As glucanases are produced for degrading fungal or yeast cell wall polymer, their concentration increased in the media. This strategy will help in increase production (Cruz et al. 1995). Similarly, numerous β-1,3-glucanases are produced by *Phanerochaete chrysosporium* on adding laminarin (glucan) in the culture media as a sole source of carbon (Kawai et al. 2006). Similarly, *Acremonium cellulolyticus* was grown on different carbon substrates like lactose, galactose, sorbose, solka flok, pullulan and cellobiose to produce cellulases (Endo-(1→4)-β-glucanases). Cellulases were produced in high amounts with the supplementation of carbon sources like Solka floc and lactose. Besides, both carbon sources acted synergistically to increase the production of cellulases (Fang et al. 2008).

Enzymes are specific to the substrate and their reaction to different glucan may vary. The degradation of glucans also depends on their structure. The difference like the hydrolysis products can be explained by the substrate specificities of each enzyme fraction, and the structural differences of the β-D-glucans attacked. For instance, two glucanases were isolated from *Trichoderma harzianum Rifai* PAMB86 and employed to degrade algal paramylon (1 → 3)-β-D-glucan) and lichen pustulan ((1 → 6)-β-D-glucan). Both glucanases acted on both glucans and produced glucose and laminaribiose as degraded products while only one glucanase could degrade pustulan and produced glucose, gentiobiose and gentio-oligosaccharide as degraded products (Giese et al. 2011).

4.3.2.3 Isolation, Purification and Characterization of the Enzyme

The desired glucanases production is induced by controlled optimum conditions (Table 4.1). The enzyme can be present either extracellularly or intracellularly. The extracellular enzyme is secreted into the media and can be isolated and purified by recovering the broth. In contrast to this, the intracellular enzyme is present in the microbial cell which is recovered by collecting the microbial biomass and breaking the cells. In this way, cell disruption techniques are required for the recovery of enzymes.

Purification techniques are focused on obtaining the enzymes in pure form. Post-production purification of enzymes from the broth or cellular biomass involves several steps like ammonium sulphate purification, gel filtration and ion-exchange chromatography. The first step of purification includes ammonium sulphate precipitation of enzymes and further, purified by the chromatographic procedure. For instance, ion-exchange chromatography using DEAE-Sephadex, gel permeation chromatography using Bio-Gel P-200 (Ghosh et al. 1980) and gel filtration chromatography on Sephacryl S-200 (Sena et al. 2011). It is pertinent to note that purification techniques must ensure minimum loss of enzymes and their activity.

Endo-(1 → 3)-β-D-glucanase (EC 3.2.1.6) (mol. Wt. 30,000) was purified by Miyazaki et al. (1977) using ammonium sulphate fractionation, gel filtration and column chromatography. The optimum conditions for the maximum enzymatic reaction were studied. The enzyme was reported to be active at pH 5, 50 °C temperature, inactivation by lead copper and mercury ions, and inhibited by zinc and iron ions.

Noronha and Ulhoa (2000) purified extracellular β-1,3- glucanases produced by *T. harzianum* growing on chitin containing medium. This was confirmed as an endoglucase type of enzyme due to the production of short oligosaccharides rather than glucose from laminarin.

4.3.3 Applications of Glucanases

Recent advancements in recombinant DNA technology and enzyme engineering, a variety of enzymes with diverse applications have been developed for their application in food, cosmetics production, therapeutics and diagnostics and fermentation industry. For different applications, enzymes are applied in different ways: (a) Direct application of enzymes and (b) immobilization of enzymes.

Enzymes can be applied directly for their function and application. However, this leads to the wastage of enzymes as it is difficult to recover all enzymes after the product recovery and their use. There are some other drawbacks like their instability

Table 4.1 Glucanases production and the optimum conditions required by different fungi

S. No.	Enzyme	Produced by	Effect of factors (Optimum conditions)			References
			pH	Temperature (°C)	Others	
1	Glucanase three different isoforms (GLUI, GLU II and GLUIII) (thermostable up to 90 °C)	*Moniliophthora perniciosa*	5.0	40	Increased by KCl and 0.2 M NaCl addition	Sena et al. (2011)
2	Endo-(1→ 6)-β-D-glucanase (EC 3.2.1)	*Mucor hiemalis*	4.5	55	Tolerance to a wide range of temperature	Miyazaki and oikawa (1976)
3	Extracellular β-1,3-glucanases (Mol.wt. 83.1 kDa.)	*Trichoderma asperellum*	5.1	55	Strongly inhibited by Hg^{2+} (mercury) ions, induced production by *Rhizoctonia solani*	Bara et al. (2003)
4	1,4-β-D-glucan 4-glucanhydrolase (EC 3.2.1.73) Mol.wt. 33.7 kDa	*Rhizopus microsporus*	4–5	50–60	Chitin addition	Celestino et al. (2006)
5	Extracellular endoglucanases (RCE1 and RCE2) (sequence homology to *Humicola insolens)*	*Rhizopus oryzae*	5–6	55	Inhibited by Cu^{2+}, Zn^{2+}, Co^{2+} and Pb^{2+}	Murashima et al. (2002)
6	β-1,3-glucanase (Mol. wt. 29 kDa)	*Trichoderma harzianum*	4.4	50 strongly inhibited by $HgCl_2$ and SDS	Strongly inhibited by $HgCl_2$ and SD	Noronha and Ulhoa (2000)
7	Endo-β-1,3-glucanase	*Aspergillus fumigatus*	5.0–6.5	24–40	Carboxymethyl-Curdlan-Remazol brilliant blue (CM-Curdlan-RBB) as a substrate	Hartl et al. (2011)
8	Endo-(1→4)-β-glucanase	*Aspergillus japonicus*	4.7	50	Highly active against barley β-glucan and lichenan (263 and 267 U/mg protein) and had much lower activity towards carboxymethylcellulose (3.9 U/mg).	Grishutin et al. (2005)

(continued)

Table 4.1 (continued)

S. No.	Enzyme	Produced by	Effect of factors (Optimum conditions)			References
			pH	Temperature (°C)	Others	
9	β-(1 → 3)-Glucanases	*Trichoderma harzianum Rifai* PAMB-86	5.5	55	Aeration of 1.5vvm	Giese et al. (2011)
10	β-1,3-glucanase (36 kDa)	*Trichoderma harzianum*	5.0	50	Chitin as sole source of carbon, insensitive to organic compound and metal ions, except for the ferric ion	Marco and Felix (2007)
11	Extracellular b-1,3-glucanase	*Trichoderma harzianum Rifai*		Agaricus blazei (Agaricus brasiliensis) extract as a substrate in submerged fermentation	Maximum activity (0.77 U/mL) in 1 day	Carneiro et al. (2011)
12	β-Glucanases	*Penicillium oxalicum* and *Penicillium citrinum*		50–55	Rice bran homogenized with 0.5% (w/v) $(NH_4)_2SO_4$ solution; no effect of Zn^{2+}; Cu^{2+} had little effect and Hg^{2+} inhibited enzymatic activity	Hamuel (2011)
13	Alpha-1,3-glucanases (molecular mass 63.8 kDa)	*Trichoderma harzianum*	3.5–4.5	50–55	–	Fuglsang et al. (2000)
14	Alpha-1,3-glucanases (63.6 kDa)	Penicillium purpurogenum	4.6	45	–	Shalom et al. (2008)

to harsh operational conditions, and inhibition due to high concentration of reactants. These problems can be resolved economically by enzyme immobilization techniques which do not affect the properties and catalytic activities of the enzymes but gives better results in terms of enzyme recovery and reuse.

Enzymes can be immobilized in different ways like adsorption, cross-linking, entrapment, covalent binding. Adsorption can be achieved by any inert organic or inorganic material such as glass beads, silica gel, etc. The enzymes can further desorb by changing pH, temperature or ionic strength. Another method of immobilization is based on the physical entrapment of enzymes in a trap. The trap is a matrix of an inert material like gelatine, starch collagen, cellulose, etc. Further, enzymes can also be immobilized by cross-linking and covalent bonding. Enzymes can be linked to supporting chemicals by cross-linkage formation like cellulose and glyceraldehyde. The drawback is associated with the denaturation of the enzyme during cross-linkage formation. However, this problem can be solved by the covalent binding method due to the covalent bonding of the enzyme to support material.

4.3.3.1 Application in the Hydrolysis of Lignocellulosic Biomass

The industrial application of lignocellulosic biomass depends on its digestion or enzymatic hydrolysis to produce fermentable sugars. Cellulases play a major role in the hydrolysis of biomass containing glucan. Inoue et al. (2014) reported the role of five enzymes and their synergistic action in the hydrolysis of acid pretreated corn stover by *Talaromyces cellulolyticus* (previously known as *Acremonium cellulolyticus*) (Inoue et al. 2014).

4.3.3.2 Application in the Brewing Industry

β-glucans are present in barley grains and their partial hydrolysis occurs at the seed germination stage by endogenous 1,3-1,4-β-glucanases. However, their inactivation at high temperature is the cause of viscosity and turbidity of the brewer's mash which further results in blocking or choking of pumps and filters. The filtration time and viscosity of brewer mash can be improved by the treatment of β-glucanases, derived from *Rhizopus microsporus* var. *microsporus* (Celestino et al. 2006). The isolation and production of 1,3-1,4-β-glucanases from *Rhizomucor miehei* and further, their application in the malting process in the brewing industry were also reported by Yang et al. (2019). The role of mixed β-glucanases and xylanases from *Humicola insolens* in the mashing process of brewing was also reported (Smith 2004) and patented.

4.4 Glucanases of Yeast

Glucanases are found to be produced by different yeasts. Brock (1965) reported that a specific type of intracellular β-glucanase can be misinterpreted for β-glucosidase due to its ability of hydrolyzing p-nitrophenyl-β-D-glucoside. However, it is not able to hydrolyse aryl and alkyl glucosides and therefore, considered as glucanases.

Chambers et al. (1993) reported the production of exo-β-(1,3)-glucanases from a yeast *Candida albicans* ATCC 10261. Nucleotide sequence showed 58% identity and 85% similarity to that of *Saccharomyces cerevisiae*. Gene was reported to be heterozygous for this glucanase which means the presence of only one copy in the diploid genome. Cutfield et al. (1999) also reported the production of exo-β-(1,3)-glucanases from *Candida albicans*.

4.4.1 Role of Yeast Glucanases

Cutfield et al. (1999) reported the role of exo-β-(1,3)-glucanases in cell wall glucan metabolism and morphogenesis of *Candida albicans*. In another study, conducted by Xu et al. (2013), the filamentation of *Candida albicans* by β-1,3-glucanase at 22–37 °C was reported.

4.4.1.1 Role of Yeast Glucanases in Establishing Infection

Certain fungi are dimorphic and grow in two morphological forms, i.e. yeast form and mycelial forms. Although filamentation of yeast is controlled by the signalling process, the role of glucanases cannot be ignored. In yeasts, the morphological transformation from yeast to multicellular fungal form is carried out by glucanases. At first, glucanases dissolve the cell wall of yeast which helps in the germination of hyphae. This hypha continues to grow and form mycelial filaments. The interwoven filamentous growth leads to the formation of biofilm. Therefore, glucanases can also induce the formation of biofilm in signal-controlled diverse environments. This helps the yeast to cause the infection in the diverse host conditions.

4.4.1.2 Role of Yeast Glucanases as a Biocontrol Agent

Glucanases from the yeasts are also acting as biocontrol agents. β-1,3-glucanase from cultured *Wickerhamomyces anomalus* EG2 was reported to inhibit the growth of two fungi, i.e. *Fusarium oxysporum* KACC 40032 and *Rhizoctonia solani* KACC 40111 (Hong et al. 2017a).

4.4.1.3 Role in Maintaining Turgor Pressure During Growth of Yeast

Alpha-1,3-glucanases, in yeast *Schizosaccharomyces pombe*, are required for secondary septum formation and maintaining the strength of the primary septum as turgor pressure increases during the septa formation and separation of the cell. In this process, alpha-glucans provide support to tolerate turgor pressure during cell separation (Hochstenbach et al. 1998).

4.4.2 *Production of Yeast Glucanases*

The yeast glucanases have several applications and therefore, it needs to be produced in high amount to use as a biocontrol agent. The fundamental process of production of yeast glucanases also depends upon the media and favourable fermentation conditions, like the production of other fungi. However, there may be a change in the culture and substrate. Culture conditions also must be adjusted according to the growth and production requirement of yeast. For instance, *Wickerhamomyces anomalus* EG2 was induced to produce β-1,3-glucanases when laminarin was used as a substrate in the media. In contrast to this, the addition of chitin in the media promotes the release of chitinase enzymes (Hong et al. 2017a). In another experiment, W. anomalus EG2 was grown in potato dextrose broth and yeast extract-peptone-dextrose media with and without chitin. Chitin suppressed the activity of glucanases. Maximum 17.34 units/mg protein and 15 units/mg protein were produced in potato dextrose broth with maximum of 0.5% colloidal chitin and yeast extract-peptone-dextrose media with 0.3% colloidal chitin (Hong et al. 2017b). An extracellular β-glucanase, i.e. endo-β-1,3(4)-glucanase, was reported to be produced by *Kluyveromyces marxianus* at pH 5.5, temperature 35 °C in the presence of 4% glucose concentration (Lopes et al. 2014).

After the production, intracellular yeast β-glucanases can be recovered by the lysis of the cell.

The cell lysate can be assessed to produce glucanases. For example, cell lysate obtained from the lysis of *Saccharomyces cerevisiae, Fabospora fragilis* and *Hansenula anomala cleaved laminarin and pustulan into glucose units. Fabospora fragilis* and *Hansenula anomala produced seven-fold higher activity as compared to Saccharomyces cerevisiae (*Abd-El-Al and Phaff 1968*).*

4.5 Glucanases of Mushroom

Mushroom produces a variety of enzymes for the degradation of lignocellulosic wastes and their growth. A novel endo-type β-1,3-glucanase (GLU1) was reported to be produced by *Lentinula edodes* which was categorized in a novel category glycoside hydrolase family GH128 (Sakamoto et al. 2005). A wide range of glucanases is produced by different mushroom species according to the media and cultivation conditions. For instance, *Trametes hirsuta* produces many types of glucanases when grown on lignocellulosic materials. Many of these enzymes were cellobiases and xylanases; however, there was a drastic change in the profile of enzyme expression. This change was also observed when the cultivation of mushrooms was transferred from submerged cultivation to surface cultivation. Availability of lignocellulosic substrates induced the secretion of endo-1,3-glucanases (Vasina et al. 2016). Similarly, Trametes hirsuta produced different enzymes including glucanases for the complete degradation of lignocellulosic substrates. When grown on soluble cellulose and filter paper, cellulases (glucanases) were reported to be produced by *Cerrena meyenii, Phaeotrametes decipiens, Trametes modesta* and *T. pocas* in high amount irrespective of the cellulosic substrates (Mswaka and Magan 1998).

The production of cellulases (glucanases) was also reported for the *Trametes hirsuta* when grown in carboxymethyl cellulose containing media. The enzyme revealed the homology with *Irpex lacteus*. The gene for coding this was isolated for a detailed study, which revealed an open reading frame of 384 amino acids (Nozaki et al. 2007).

4.5.1 Role of Mushroom Glucanases

Mushroom glucanases play a different role in their growth. These also control the process of autolysis. There are many other functions of mushroom glucanases that are described here.

4.5.1.1 Role of Mushroom Glucanases in Autolysis of Fruit Body

Mushrooms undergo autolysis after the completion of their life cycle. As the fungal cell wall is made up of glucans, glucanases play an important role in degrading the cell wall and cause the lysis of mushroom fruit bodies. In *Coprinopsis cinerea* fruiting bodies, the cell wall is made up of a 1,3-β-glucan backbone. The enzymes endo-1,3-β-glucanase, exo-1,3-β-glucanase and 1,3-β-glucosidase acted synergistically to degrade the 1,3-β-glucan cell wall during fruiting body autolysis (Zhou et al. 2015). Autolysis of *Coprinopsis cinerea* fruit bodies needs to be controlled to increase its commercial value. This is possible by controlling the action of β-1,3-glucanases on the cell wall. To prove this, a mutant of *C. cinerea* was developed by using ultraviolet mutagenesis. These mutants were reported to form normal pileus but lack glucanases activity and therefore, the shelf life of pileus increased (Liu et al. 2015). All autolysis processes are controlled and genetically regulated processes. Sakamoto et al. (2005) isolated exo-β-1,3-glucanase and revealed its amino acid sequence like *Agaricus bisporus* (67% identity) and yeast exo-β-1,3-glucanases (40% identity). The gene encoding this enzyme was reported to be expressed in the stipe of mushroom rather than pileus of young fruit bodies. Besides, it is expressed in the gills of mature fruit bodies but did not express in the vegetative mycelium. In this way, the concentration of glucanases in a different part of the mushroom and different stages of development is a controlled and genetically regulated process (Sakamoto et al. 2005). Similarly, β-1,6-glucanase, LePus30A, was reported to be responsible for the hydrolysis of *Lentinula edodes* fruit bodies (Konno and Sakamoto 2011).

4.5.1.2 Role of Mushroom Glucanases as a Biocontrol Agent

Like other fungi, mushrooms and their extract can be used as a biocontrol agent. These extracts can induce the plant to produce glucanases and defence proteins and therefore, provide resistance to the pathogens. The detailed role of glucanases and mushrooms in preventing infection is mentioned in Table 4.2.

4.5.1.3 Role of Mushroom Glucanases as an Antioxidant Agent

Trametes versicolor produces polysaccharopeptides which prevents oxidative damage due to its antioxidant activities. A short bioactive peptide was generated by treating the polysaccharopeptide with glucanases which possessed more potential to inhibit oxidative damage. These glucanase treated peptides decreased the formation of reactive oxygen species and increased the survival of the cell. The glucanases

Table 4.2 Role of different glucanases in controlling infection caused by microbes in plants

S. No.	Glucanases	Source mushroom	Plant disease/pathogen	Reference
1.	β-1,3-glucanase	Culture filtrates of *Hericium erinaceus*, *Lentinula edodes* (Sanjo 701), *Grifola frondosa* and *Hypsizygus marmoreus*	Tomato wilt bacteria *Ralstonia solanacearum* B3; bacterial wilt disease of tomato, induced expressions of plant defence genes encoding β-1,3-glucanase (GluA) and pathogenesis-related protein-1a (PR-1a),	Kwak et al. (2015)
2.	Exo-β-1,3-glucanase (EC 3.2.1.58)	*Pichia anomala* strain K	*Botrytis cinerea* on postharvest apples	Jijakli and Lepoivre (1998)
3.	β-(1,3)-glucanase	*Trametes versicolor* and *Pleurotus eryngii*	*Fusarium oxysporum* f. sp. *lycopersici* race 2 (FOL 2)	Ruiz-Dueñas and Martínez (1996)

treatment reduced the size of polysaccharopeptide without affecting its functional groups (Jhan et al. 2016).

4.5.2 Production of Mushroom Glucanases

The production of mushroom glucanases can be done in the same way as mentioned for fungi. Like fungi, the production is affected by the choice of substrate. When corncob was used as a substrate for the cultivation of *Pleurotus ostreatus* DBUI14 had the lowest endoglucanase activity. When grown on rice bran exoglucanase activity was reported to be high (Ekundayo et al. 2017).

The production of mushroom glucanases varies with the stages of development. In the initial stage, of mycelial growth, a variety of enzymes along with glucanases (both exo- and endo-β-1,4-glucanases) are produced. This is possibly due to the growing requirement of vegetative growth. This continues till mycelium covers 2/3 of the substrate. However, the activity and concentration of the enzyme dropped during the fruit bodies formation and till it reaches maturity. After harvesting the fruit bodies of the mushroom, the enzyme concentration increases to support the growth of newly grown mycelium in the second flush (Xie et al. 2016).

This proves the importance of different stages of mushroom in glucanases production. The high amount of enzyme production requires the precise selection of state in solid-state fermentation.

Glucanases are reported to produce during the storage of mushrooms. Kawakami et al. (2004) reported the effectiveness of low temperature in maintaining the lentinan, an antitumor polysaccharide, of mushroom *Lentinus edodes*. The lentinan content of stored mushrooms decreased in 7 days due to the production of glucanases but remains unaffected till 5 days at 20 °C 100% CO_2 (Kawakami et al. 2004).

The production of β-1,3-glucanases is affected by the presence of other fungi. Xing and Guo (2004) reported an increased production of glucanases by *Grifola umbellata* when induced by companion fungus. It is also associated with the sclerotia formation in *G. umbellata* (Xing and Guo 2004) (Table 4.3).

4.5.3 Applications of Mushroom Glucanases in Traditional Therapies

Mushroom β-glucan is used in several applications including medical and therapeutics. β-1,6-glucan plays an important role in the immunomodulatory activity and therefore, can be used for treating cancer. An endo-β-1,6-glucanase was isolated from *Trichoderma harzianum* and used to hydrolyse MD-fraction which was recovered from the mushroom *Grifola frondosa*. The mol. wt. of treated MD-fraction was reduced to 10 times due to the release of dextran units. When this low molecular

Table 4.3 Mushroom glucanases and their production

S. No.	Enzyme	Produced by	Effect of factors			Reference
			pH	Temperature (°C)	Others	
1	Extracellular β-glucanase, i.e. exo-β-1,3-1,6-glucanase	*Flammulina velutipes* (Enoki)	6.1	60	Inactivated by 1 mM mercuric ions.	Fukuda et al. (2008)
2	1,3-β-glucanase	*Pleurotus ostreatus* DBUI14 and *P. pulmonarius* DBUI002	4.8	50	Rice bran, corn cob and sawdust as substrates production in 15 days	Ekundayo et al. (2017)
3	1,3-β-D-glucanases	*Ganoderma tsugae*	5	50	Produced against laminarin, increased rapidly after the stationary growth phase	Liang et al. (1995)

weight fraction is introduced by intraperitoneal infection, carcinoma is suppressed. This might be possible due to the increased concentration of cytokines. This showed the antitumor potential of β-glucanase treated low molecular weight fraction (MD-fraction) of *Grifola frondosa* (Miyake et al. 2006).

The above-mentioned study revealed the importance of β-glucanase activity in fungi and its application in treating cancer. Therefore, it is of utmost importance to find out the glucanases content of the mushroom fruit bodies. Recently, Yamanaka et al. (2020) conducted an ELISA-like assay to find out the presence of β-1,3-glucan and β-1,6-glucan in mushrooms using modified β-1,6-glucanase and soluble dectin-1-Fc as the probes. In this case, crude hot water extracts of edible mushrooms (*Grifola frondosa, Agaricus bisporus, Pleurotus tuoliensis, P. eryngii, P. ostreatus, Hypsizygus marmoreus* and *Lentinus edodes*) were tested for the presence of β-glucans. This study revealed the highest glucan amount in *L. edodes* which showed its importance in treating cancer (Yamanaka et al. 2020).

4.6 Conclusion

This chapter deals with the current and traditional knowledge and the role of different glucanases. Besides, these glucanases can be applied as a biocontrol agent to inhibit the growth of pathogens. These are also useful in the fungal morphogenesis of yeast, multicellular fungi as well as mushrooms. These can be produced at a large scale under controlled environmental conditions. A recent application in cancer therapy is opening a new aspect for research.

References

Abd-El-Al ATH, Phaff HJ (1968) Exo- β-glucanases in yeast. Biochem J 109(3):347–360. https://doi.org/10.1042/bj1090347

Adams DJ (2004) Fungal cell wall chitinases and glucanases. Microbiology 150(Pt 7):2029–2035. https://doi.org/10.1099/mic.0.26980-0

Amey RC, Mills PR, Bailey A, Foster GD (2003) Investigating the role of a *Verticillium fungicola* β-1,6-glucanase during infection of *Agaricus bisporus* using targeted gene disruption. Fungal Genet Biol 39(3):264–275. https://doi.org/10.1016/S1087-1845(03)00061-6

Bara MT, Lima AL, Ulhoa CJ (2003) Purification and characterization of an exo-beta-1,3-glucanase produced by *Trichoderma asperellum*. FEMS Microbiol Lett 219(1):81–85. https://doi.org/10.1016/S0378-1097(02)01191-6

Brock TD (1965) β-Glucanase of yeast. Biochem Biophys Res Commun 19(5):623–629. https://doi.org/10.1016/0006-291X(65)90385-2

Carneiro AAJ, Giese EC, Barbosa AM, Gomes E, Da Silva R (2011) B-Glucanase produced by *T. harzianum* on *a. blazei*. Food Technol Biotechnol 49(3):341–346

Celestino KR, Cunha RB, Felix CR (2006) Characterization of a beta-glucanase produced by *Rhizopus microsporus* var. *microsporus*, and its potential for application in the brewing industry. BMC Biochem 7(23). https://doi.org/10.1186/1471-2091-7-23

Chambers RS, Broughton MJ, Cannon RD, Carne A, Emerson GW, Sullivan PA (1993) An exo-beta-(1,3)-glucanase of *Candida albicans*: purification of the enzyme and molecular cloning of the gene. J Gen Microbiol 139(2):325–334. https://doi.org/10.1099/00221287-139-2-325

Cruz JDL, Pintor-toro JA, Beni'Tez T, Llobell A, Romero LC (1995) A novel Endo-β-1,3-Glucanase, BGN13.1, involved in the mycoparasitism of *Trichoderma harzianum*. J Bacteriol 177(23):6937–6945

Cutfield SM, Davies GJ, Murshudov G, Anderson BF, Moody PC, Sullivan PA, Cutfield JF (1999) The structure of the exo-beta-(1,3)-glucanase from *Candida albicans* in native and bound forms: relationship between a pocket and groove in family 5 glycosyl hydrolases. J Mol Biol 294(3): 771–783. https://doi.org/10.1006/jmbi.1999.3287

Ekundayo FO, Ekundayo EA, Ayodele BB (2017) Comparative studies on glucanases and β-glucosidase activities of *Pleurotus ostreatus* and *P. pulmonarius* in solid state fermentation. Mycosphere 8(8):1201–1209. https://doi.org/10.5943/mycosphere/8/8/16

Fang X, Yano S, Inoue H, Sawayama S (2008) Lactose enhances cellulase production by the filamentous fungus *Acremonium cellulolyticus*. J Biosci Bioeng 106(2):115–120. https://doi.org/10.1263/jbb.106.115

Fuglsang CC, Berka RM, Wahleitner JA, Kauppinen S, Shuster JR, Ramussen G, Halkier T, Dalboge H, Henrissat B (2000) Biochemical analysis of recombinant fungal mutanasese: a new family of a-1,3-glucanases with novel carbohydrate-binding domains. J Biol Chem 275: 2009–2018

Fujikawa T, Sakaguchi A, Nishizawa Y, Kouzai Y, Minami E, Yano S, Koga H, Meshi T, Nishimura M (2012) Surface a-1,3-glucan facilitates fungal stealth infection by interfering with innate immunity in plants. PLoS Pathog 8:e1002882

Fukuda K, Hiraga M, Asakuma S, Arai I, Sekikawa M, Urashima T (2008) Purification and characterization of a novel exo-beta-1,3-1,6-glucanase from the fruiting body of the edible mushroom Enoki (*Flammulina velutipes*). Biosci Biotechnol Biochem 72(12):3107–3113. https://doi.org/10.1271/bbb.80213

Ghosh AK, Banerjee PC, Sengupta S (1980) Purification and properties of xylan hydrolase from mushroom *Termitomyces clypeatus*. Biochim Biophys Acta 612(1):143–152. https://doi.org/10.1016/0005-2744(80)90287-9

Giese EC, Dekker RFH, Barbosa AM, da Silva MLC, da Silva R (2011) Production of β-(1,3)-glucanases by *Trichoderma harzianum Rifai*: optimization and application to produce gluco-oligosaccharides from paramylon and pustulan. Ferment Technol 1:1. https://doi.org/10.4172/2167-7972.1000102

Grenier J, Potvin C, Asselin A (2000) Some fungi express β-1,3-glucanases similar to thaumatin-like proteins. Mycologia 92(5):841–848. https://doi.org/10.2307/3761579

Grishutin SG, Gusakov AV, Dzedzyulya EI, Sinitsyn AP (2005) A lichenase-like family 12 endo-(1→4)-beta-glucanase from *aspergillus japonicus*: study of the substrate specificity and mode of action on beta-glucans in comparison with other glycoside hydrolases. Carbohydr Res 341(2): 218–229. https://doi.org/10.1016/j.carres.2005.11.011

Hamuel DJ (2011) Production of β-glucanase enzyme from *Penicillium oxalicum* and *Penicillium citrinum*. Afr J Biotechnol 10(47):9657–9660. https://doi.org/10.5897/AJB07.484

Hartl L, Gastebois A, Aimanianda V, Latgé JP (2011) Characterization of the GPI-anchored endo β-1,3-glucanase Eng2 of *aspergillus fumigatus*. Fungal Genet Biol 48(2):85–191. https://doi.org/10.1016/j.fgb.2010.06.011

Hochstenbach F, Klis FM, Van Den Ende H, Van Donselaar E, Peters PJ, Klausner RD (1998) Identification of a putative alpha-glucan synthase essential for cell wall construction and morphogenesis in fission yeast. Proc Natl Acad Sci U S A 95:9161–9166

Hong SH, Song YS, Seo DJ, Kim KY, Jung WJ (2017a) Antifungal activity and expression patterns of extracellular chitinase and β-1,3-glucanase in *Wickerhamomyces anomalus* EG2 treated with chitin and glucan. Microb Pathog 110:159–164. https://doi.org/10.1016/j.micpath.2017.06.038

Hong SH, Song YS, Seo DJ, Kim KY, Jung WJ (2017b) Enzyme activity and expression pattern of intra- and extracellular chitinase and β-1,3-glucanase of *Wickerhamomyces anomalus* EG2 using glycol chitin and glucan-containing. Int J Biol Macromol 105(Pt 1):410–415. https://doi.org/10.1016/j.ijbiomac.2017.07.065

Inbar J, Chet I (1997) Lectins and biocontrol. Crit Rev Biotechnol 17(1):1–20. https://doi.org/10.3109/07388559709146604

Inoue H, Decker SR, Taylor LE, Yano S, Sawayama S (2014) Identification and characterization of core cellulolytic enzymes from *Talaromyces cellulolyticus* (formerly *Acremonium cellulolyticus*) critical for hydrolysis of lignocellulosic biomass. Biotechnol Biofuels 7:151. https://doi.org/10.1186/s13068-014-0151-5

Jhan MH, Yeh CH, Tsai CC, Kao CT, Chang CK, Hsieh CW (2016) Enhancing the antioxidant ability of T*rametes versicolor* polysaccharopeptides by an enzymatic hydrolysis process. Molecules 21(9):1215. https://doi.org/10.3390/molecules21091215

Jijakli MH, Lepoivre P (1998) Characterization of an exo-β-1,3- glucanase produced by *Pichia anomala* strain K, antagonist of *Botrytis cinerea* on apples. Phytopathology 88:335–343

Kawai R, Igarashi K, Samejima M (2006) Gene cloning and heterologous expression of glycoside hydrolase family 55 beta-1,3-glucanase from the basidiomycete *Phanerochaete chrysosporium*. Biotechnol Lett 28(6):365–371. https://doi.org/10.1007/s10529-005-6179-7

Kawakami S, Minato KI, Tokimoto K, Fujitake N, Mizuno M (2004) Changes of Lentinan contents and glucanase activity in *Lentinus edodes* (Berk.) singer (Agaricomycetideae) stored under controlled atmosphere. International Journal of Medicinal Mushrooms 6(1):6. https://doi.org/10.1615/IntJMedMushr.v6.i1.60

Konno N, Sakamoto Y (2011) An endo-β-1,6-glucanase involved in *Lentinula edodes* fruiting body autolysis. Appl Microbiol Biotechnol 91:1365–1373. https://doi.org/10.1007/s00253-011-3295-2

Kwak AM, Min KJ, Lee SY, Kang HW (2015) Water extract from spent mushroom substrate of *Hericium erinaceus* suppresses bacterial wilt disease of tomato. Mycobiology 43(3):311. https://doi.org/10.5941/MYCO.2015.43.3.311

Liang ZC, Hseu RS, Wang HH (1995) Partial purification and characterization of a 1,3-β-D-glucanase from *Ganoderma tsugae*. J Ind Microbiol 14:5–9. https://doi.org/10.1007/BF01570058

Liu Z, Niu X, Wang J, Zhang W, Yang M, Liu C, Xiong Y, Zhao Y, Pei S, Qin Q, Zhang Y, Yu Y, Yuan S (2015) Comparative study of non-autolytic mutant and wild-type strains of *Coprinopsis cinerea* supports an important role of glucanases in fruiting body autolysis. J Agric Food Chem 63(43):9609–9614. https://doi.org/10.1021/acs.jafc.5b03962

Lopes MR, de Souza CJ, Rodrigues MQ, Costa DA, dos Santos AF, de Oliveira LL, Ramos HJ, Guimarães VM, Silveira WB, Passos FM, Fietto LG (2014) Production and characterization of β-glucanase secreted by the yeast *Kluyveromyces marxianus*. Appl Biochem Biotechnol 172(5): 2412–2424. https://doi.org/10.1007/s12010-013-0683-3

Marco JL, Felix CR (2007) Purification and characterization of a beta-glucanase produced by *Trichoderma harzianum* showing biocontrol potential. Braz Arch Biol Technol 50(1):21–29. https://doi.org/10.1590/S1516-89132007000100003

Martin K, McDougall BM, McIlroy S, Chen J, Seviour RJ (2007) Biochemistry and molecular biology of exocellular fungal beta-(1,3)- and beta-(1,6)-glucanases. FEMS Microbiol Rev 31(2): 168–192. https://doi.org/10.1111/j.1574-6976.2006.00055.x

Millet N, Moya-Nilges M, Sachse M, Krijnse Locker J, Latgé JP, Mouyna I (2019) *Aspergillus fumigatus* exoβ(1-3)glucanases family GH55 are essential for conidial cell wall morphogenesis. Cell Microbiol 21(12):e13102. https://doi.org/10.1111/cmi.13102

Miyake N, Masuda Y, Murata Y, Inoue M, Matuda K, Okamoto K, Nanba H, Kodama N (2006) Low molecular weight maitake MD-fraction (Klasma-MD) hydrolyzed with endo-β-1,6-glucanase of *Trichoderma harzianum* induces antitumor activities. Mycoscience 47:78–83. https://doi.org/10.1007/s10267-005-0274-4

Miyazaki T, Yadomae T, Yamada H, Oikawa N (1977) An endo-(1→3)-β-d-glucanase from *Mucor hiemalis*. Carbohydr Res 55(1):65–72

Miyazaki T, Oikawa N (1976) An endo-(1→6-β-D-glucanase) from *Mucor hiemalis*. Carbohydr Res 48(2):209–216. https://doi.org/10.1016/S0008-6215(00)83216-4

Mouyna I, Hartl L, Latgé JP (2013) β-1,3-glucan modifying enzymes in *aspergillus fumigatus*. Front Microbiol 4:81. https://doi.org/10.3389/fmicb.2013.00081

Moy M, Li HM, Sullivan R, White JF Jr, Belanger FC (2002) Endophytic fungal β-1,6-glucanase expression in the infected host grass. Plant Physiol 130:1–11

Mswaka AY, Magan N (1998) Wood degradation, and cellulase and ligninase production by *Trametes* and other wood-inhabiting basidiomycetes from indigenous forests of Zimbabwe. Mycol Res 102(11):1399–1404. https://doi.org/10.1017/S0953756298006789

Murashima K, Nishimura T, Nakamura Y (2002) Purification and characterization of new endo-1,4-β-D-glucanases from *Rhizopus oryzae*. Enzym Microb Technol 30(3):319–326. https://doi.org/10.1016/S0141-0229(01)00513-0

Noronha EF, Ulhoa CJ (2000) Characterization of a 29-kDa β-1,3-glucanase from *Trichoderma harzianum*. FEMS Microbiol Lett 183(1):119–123. https://doi.org/10.1111/j.1574-6968.2000.tb08944.x

Nozaki K, Seki T, Matsui K, Mizuno M, Kanda T, Amano Y (2007) Structure and characteristics of an endo-β-1,4-glucanase, isolated from Trametes hirsuta, with high degradation to crystalline cellulose. Biosci Biotechnol Biochem 71(10):2375–2382. https://doi.org/10.1271/bbb.60385

Prasetyawan S, Sulistyowati L, Aulanni'am (2017) Glucanase and chitinase from some isolates of endophytic fungus *Trichoderma* spp. IOP Conference Series: Materials Science and Engineering, Vol. 299, International Conference on Chemistry and Material Science (IC2MS) 2017 4–5 November 2017, Malang, East Java, Indonesia

Reese ET, Parrish FW, Mary M (1962) β-d-1,6-GLUCANASES IN FUNGI. Can J Microbiol 8(3): 327–334. https://doi.org/10.1139/m62-045

Ruiz-Dueñas F, Martínez M (1996) Enzymatic activities of *Trametes versicolor and Pleurotus eryngii* implicated in biocontrol of *fusarium oxysporum* f. sp. *lycopersici*. Curr Microbiol 32: 151–155. https://doi.org/10.1007/s002849900027

Sakamoto Y, Irie T, Sato T (2005) Isolation and characterization of a fruiting body-specific exo-beta-1,3-glucanase-encoding gene, exg1, from *Lentinula edodes*. Curr Genet 47(4): 244–252. https://doi.org/10.1007/s00294-005-0563-7

Sena AR, Júnior GLV, Neto AG (2011) Production, purification and characterization of a thermostable β-1,3-glucanase (laminarinase) produced by Moniliophthora perniciosa Anais da Academia Brasileira de Ciências 83(2). Rio de Janeiro June 2011 https://doi.org/10.1590/S0001-37652011005000007

Shalom G, Pratten J, Wilson M, Nair SP (2008) Cloning heterologous gene expression and biochemical characterization of the alpha 1,3-glucanase from the filamentous fungus *Penicillium purpuro genum*. Protein Expr Purif 60(2):170–175

Smith J (2004) Mixed β-glucanase, xylanase from *Humicola insolens* chemical and technical assessment (CTA) first draft, 61st JECFA

Suyotha W, Yano S, Wakayama M (2016) α-1,3-Glucanase: present situation and prospect of research. World J Microbiol Biotechnol 32(2):30. https://doi.org/10.1007/s11274-015-1977-0

Vasina DV, Pavlov AR, Koroleva OV (2016) Extracellular proteins of *Trametes hirsuta* st. 072 induced by copper ions and a lignocellulose substrate. BMC Microbiol 16(1):106. https://doi.org/10.1186/s12866-016-0729-0

Xie C, Yan L, Gong W, Zhu Z, Tan S, Chen D, Hu Z, Peng Y (2016) Effects of different substrates on lignocellulosic enzyme expression, enzyme activity, substrate utilization and biological efficiency of *Pleurotus Eryngii*. Cell Physiol Biochem 39:1479–1494. https://doi.org/10.1159/000447851

Xing XK, Guo SX (2004) Effects of companion fungus on several enzymatic activities of *Grofola umbellate*. Zhongguo Zhong Yao Za Zhi 29(4):310–313

Xu H, Nobile CJ, Dongari-Bagtzoglou A (2013) Glucanase induces filamentation of the fungal pathogen *Candida albicans*. PLoS One 8(5):e63736. https://doi.org/10.1371/journal.pone.0063736

Yamanaka D, Ishibashi KI, Adachi Y, Ohno N (2020) Quantification of mushroom-derived soluble β-1,6-glucan using the function-modified recombinant β-1,6-glucanase. Int J Med Mushrooms 22(9):855–868. https://doi.org/10.1615/IntJMedMushrooms.2020035888

Yang Y, Zhang H, Zuo J, Gong X, Yi F, Zhu W, Li L (2019) Advances in research on the active constituents and physiological effects of *Ganoderma lucidum*. Biomed Dermatol 3:6. https://doi.org/10.1186/s41702-019-0044-0

Yeoh KA, Othman A, Meon S, Abdullah F, Ho CL (2012) Sequence analysis and gene expression of putative exo- and endo-glucanases from oil palm (*Elaeis guineensis*) during fungal infection. J Plant Physiol 169(15):1565–1570. https://doi.org/10.1016/j.jplph.2012.07.006

Zhou Y, Zhang W, Liu Z, Wang J, Yuan S (2015) Purification, characterization and synergism in autolysis of a group of 1,3-β-glucan hydrolases from the pilei of *Coprinopsis cinerea* fruiting bodies. Microbiology t 161(10):1978–1989. https://doi.org/10.1099/mic.0.000143

Zonneveld BJ (1972) Morphogenesis in *aspergillus nidulans*. The significance of an alpha-1, 3-glucan of the cell wall and alpha-1,3-glucans for cleistothecium development. Biochim Biophys Acta 273:174–187

Chapter 5
Beta-Glucanases in Animal Nutrition

Lekshmi K. Edison, V. M. Ragitha, and N. S. Pradeep

Abstract Many nutrients such as proteins, lipids, and starch can be hydrolysed in the digestive tracts of animals. Although, due to a lack of hydrolysing enzymes, the majority of non-starch polysaccharides, such as beta-glucan, remain undigested. Beta-glucan rich cereals such as barley, oats, and wheat are used in animal feeds to boost immunological function, act as an alternative to antibiotics and improve the overall growth performances of animals. Apart from its positive effects, beta-glucan causes some negative effects in most animals and poultry. Beta-glucan content in animal feeds causes high digesta viscosity, sequentially it reduces nutrient digestibility and disruption of intestinal inhabitant microbiota. This can be overcome by the proper digestion of beta-glucan accomplished by the supplementation of exogenous beta-glucanase in animal feeds. Exogenous beta-glucanases efficiently eradicate the negative effects associated with beta-glucan and maintain the natural microbiota in the gastrointestinal tract. This chapter explains a more comprehensive understanding of the effect of beta-glucans in animal diets and the benefit of beta-glucanases as a feed enzyme.

Keywords Feed enzymes · Benefits · Beta-glucan · Beta-glucanases · Prebiotics · Intestinal microbiota

L. K. Edison
Rajiv Gandhi Centre for Biotechnology (RGCB), Thiruvananthapuram, Kerala, India

V. M. Ragitha
KSCSTE-Jawaharlal Nehru Tropical Botanic Garden and Research Institute, Thiruvananthapuram, Kerala, India

N. S. Pradeep (✉)
KSCSTE-Malabar Botanical Garden and Institute for Plant Sciences, Kozhikode, Kerala, India

N. S. Pradeep, L. K. Edison (eds.), *Microbial Beta Glucanases*, Interdisciplinary Biotechnological Advances, https://doi.org/10.1007/978-981-19-6466-4_5

5.1 Introduction

Enzymes are natural catalysts produced by living organisms to accelerate all the biochemical reactions by reducing activation energy. The three-dimensional structure and amino acid positions determine the catalytic efficiency of enzymes. Most of the enzymes are very sensitive to environmental changes, and they need few optimum conditions for the proper functioning and efficient catalysis (Koolman and Roehm 2005). In animal and poultry farming, the major concern of farmers is the cost and nutritional value of feed. The limiting factor is their capability to hydrolyse fibres. The addition of certain enzymes can overcome the digestibility problems of animal feed, thereby enhances the nutritional turnover (Bedford and Partridge 2001). The addition of exogenous feed enzymes eliminates the fibres and anti-nutritional factors which are not accessible by endogenous enzymes and helps to improve the digestion and proper dissemination of other nutritional elements. The early research on feed enzymes has been conducted in the United States, which studied the nutritional aspects of enzyme preparations with barley-based diets. The enzymes evaluated in these studies were small quantities of crude enzymes (Ewing 1963; Jensen et al. 1957; Fry et al. 1958; Willingham et al. 1961). The feed enzyme market is a rapidly growing commercial market.

Animal feed ingredients are classified mainly into three types based on the approachability to feed enzymes: (1) ingredients that can digest by animals using endogenous enzymes (e.g. proteins, lipids, and starch), (2) ingredients that cannot be digested by endogenous enzymes (e.g. cellulose), (3) ingredients that cannot be digested by endogenous enzymes and also have possess anti-nutritive effects (e.g. pentosans, β-glucans, phytate). The use of exogenous feed enzymes improves the digestion of the last two ingredient categories. The nutrients and anti-nutrients in feed exist as complexes that reduces their accessibility to feed enzymes. However, the information regarding the chemistry and structure of the ingredients increases the usefulness of feed enzymes, because the chemistry of each ingredient is different. Another factor is the usage of the suitable enzyme to substrate ratio in the feeds (Choct et al. 2004). Currently, numerous commercial feed enzymes from different sources are available. Each shows varying enzyme activities and optimal characteristics. Mainly four categories are commercially available (Fig. 5.1).

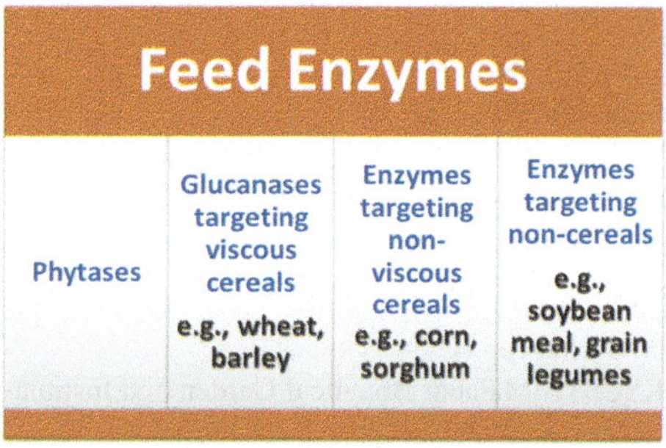

Fig. 5.1 Classification of animal feed enzymes

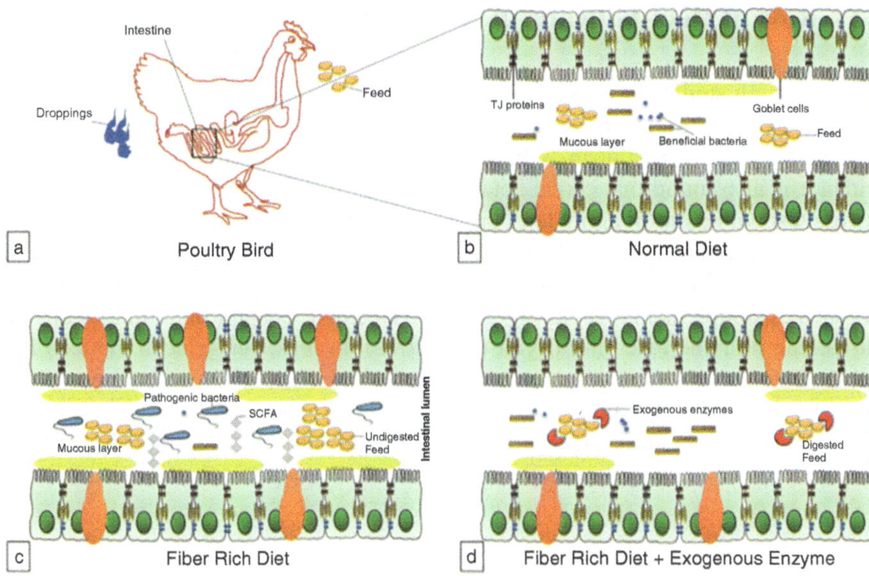

Fig. 5.2 Effect of fibre (e.g. beta-glucan) rich diet, exogenous feed enzymes, and intestinal health. (**a**) poultry bird, (**b**) normal intestinal lumen with beneficial microorganisms, (**c**) intestinal lumen with high viscous environment increases mucous, undigested feed, pathogenic bacteria, short-chain fatty acids, (**d**) intestinal lumen containing exogenous enzymes replaces normal physiology, beneficial bacteria and improves feed digestion (Raza et al. 2019)

Most of the feed enzymes are available as a cocktail mixture of two or more enzymes, except phytases. All nutrients in foodstuff exist as complexes. Hence, multiple enzyme strategy in animal diet delivers a synergic effect that enhances the utilisation of nutrients in feedstuffs (Cowieson et al. 2006; Selle and Ravindran 2007).

The digestion efficiency is controlled by the physiological conditions of the stomach. The ultimate role of feed enzymes is to improve the digestion of dietary components. The use of enzymes improves the digestion efficiency of the stomach up to 100%. In addition, there are many other justifications for the widespread use of feed enzymes. It increases the choice and flexibility in feed formulations, reduces variability between good- and poor-quality samples, reduces moisture content, lowers wet litter in poultry feeds with high levels of non-starch polysaccharides (Collett 2012), improves gut health by shifting the gut flora with beneficial microbial species (Fig. 5.2) (Bedford and Cowieson 2012; Yegani and Korver 2008), improves intestinal morphology, which enhances better nutrient absorption (Viveros et al. 1994; Jaroni et al. 1999), lowers manure output, maintains uniformity of animals in the market by increasing the performance of growth.

5.2 Role of Beta-Glucans in Animal Immunity

Many research studies have been conducted to prove the impact of fungi and yeast beta-glucan on immunity in animals (Murphy et al. 2010; de Oliveira et al. 2019). Though the yeast beta-glucan is found to be an effective immunostimulant, some evidence is also found that cereal beta-glucan can improve the immune system and gastrointestinal health in animals (Yun et al. 2003; Estrada et al. 1997). Noss et al. 2013, confirmed beta 1,3–1,4beta-glucan from cereal grains can induce innate immunity. Cereal beta-glucans recognise and bind to the pattern recognition receptors such as complement receptor 3 (CR3) and Dectin-1 receptor on immune cells for stimulating immune reaction (Taylor et al. 2002). In pigs, beta-glucan induced NK cell-mediated cytotoxicity by stimulating porcine monocytes and triggered the secretion of IL-10. This helps to the killing of malignant and virus-infected cells (Fig. 5.3) (Hermans et al. 2021). Studies indicated that oats beta-glucan activated human dendritic cells by binding through Dectin-1 receptors (Sahasrabudhe et al.

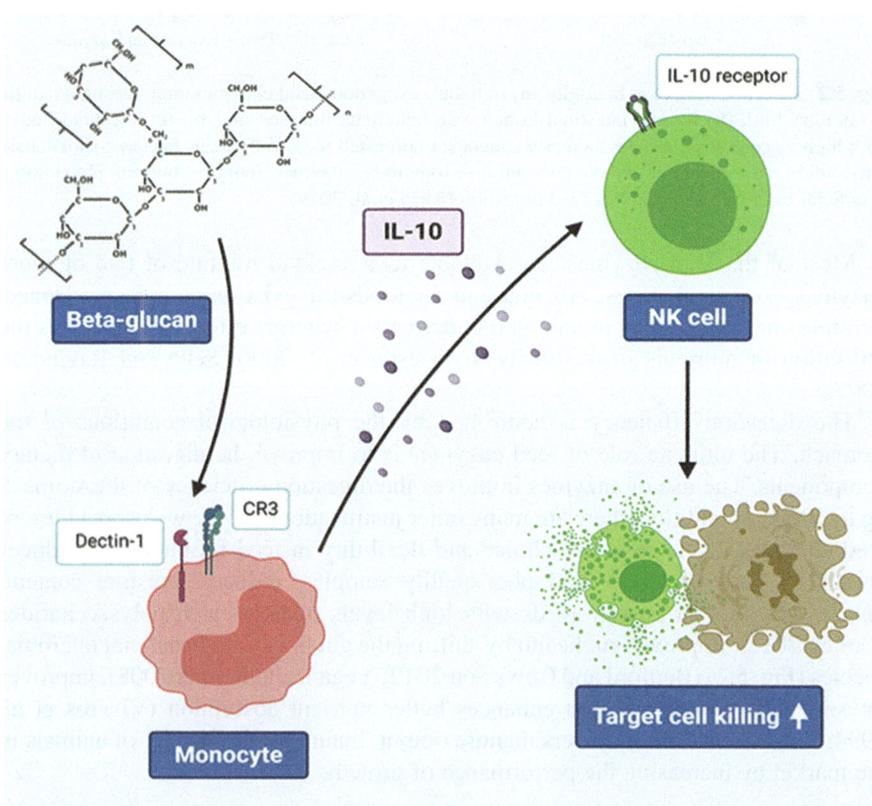

Fig. 5.3 Beta-glucan induced IL-10 secretion by monocytes triggers porcine NK cell cytotoxicity (Hermans et al. 2021)

2016). The confirmation, solubility, and molecular size of beta-glucan molecules influence their role in immune-modulating properties.

The use of formulating feed is an increasing trend to discreate the use of growth promoters (Langhout 2000; Mellor 2000). Beta-glucans play an important role to improve body weight gain in poultry. 0.1% dietary beta-glucans in the diet improves the relative weight of the spleen and the bursa of Fabricius (Cho et al. 2013). Beta-glucan also showed an improvement in thymus weight (Morales-López et al. 2009) and increases red blood cells count in broilers (Guo et al. 2003).

5.3 Beta-Glucans in Animal Diets

Barley and oats are the major sources of beta-glucans in animal feeds. Barley and oats feeding harmfully distress the digestive physiology of poultry animals by increasing digesta viscosity (Jacob and Pescatore 2014). High digesta viscosity affects the hydrolysis of protein, fat, starch, etc. and results in poor digestion and reduced nutrient absorption (Mathlouthi et al. 2002a; Rodríguez et al. 2012). The reason for the reduction in digestibility is the altered physicochemical nature of the small intestine by high digesta viscosity. Stomach and intestinal digesta viscosity reduce the interaction of digestive enzymes with substrates, reduce nutrient diffusion rate and absorption. High stomach viscosity mostly affects water-insoluble fat digestion and absorption (Campbell et al. 1983). An increase in viscosity and the low digestibility of nutrients are directly or indirectly related to the size of the digestive tract. Barley and oats feed increases excreta moisture that negatively affects the environment and sanitation in poultry barns (Mathlouthi et al. 2002b; Kaldhusdal and Hofshagen 1992). High beta-glucan content in the small intestine increases the multiplication of anaerobic microorganisms (Raza et al. 2019). Intestinal microbiome imbalance causes de-conjugation and reduced circulation of bile salts which negatively affects the digestion of fats (Feighner and Dashkevicz 1988). A high load of pathogenic bacteria also causes the initiation of enteric diseases like necrotic enteritis in animals (Annett et al. 2002; Timbermont et al. 2011).

In pigs also, beta-glucan considered as an anti-nutrient factor because it harmfully affects pig production. Like in chickens, it causes to increase digesta viscosity, reduces nutrient digestibility, absorption, and performance (Hooda et al. 2011). Due to the significant alterations in digestive tract physiology, the rate of viscosity is much lower in pigs than in poultry (Bach Knudsen and Hansen 1991; Inborr et al. 1993). Ruminants are easily able to digest the barley beta-glucan. Beta-glucan hydrolysis in beef cattle is 98.6% (Engstrom et al. 1992). Hence, digestibility and absorption of other nutrients are not affected by the beta-glucan. The beta-glucanase activity of microorganisms in the ruminant intestine and the fermentation of beta-glucan in the ruminant stomach is very high. This makes the digestion of beta-glucan easier in ruminants (Teather and Wood 1982).

The addition of oats and barley like fibres rich in cereals in single-stomached animal feed has great emphasis, as they reduce the use of feed antibiotics. Previous research has shown that major cell wall components such as endosperm and aleurone layer have changed the nutritional impacts in animal feeds. The major cell wall component is water-soluble beta-glucan, a carbohydrate made up of glucose units. Apart from cereal grains, it is also found in yeast, algae, and fungi. Depending on the source and origin, beta-glucan structure is varying (Vetvicka and Vetvickova 2014; Synytsya and Novak 2014). The structure subsequently affects the functionality of the molecule. Cereal beta-glucans are unbranched, contain beta-1,4 and beta-1,3 glycosidic linkages. Yeast and fungal beta-glucans are differing from cereal beta-glucans, it contains beta-1,6 linkages as side chains (Schwartz and Vetvicka 2021). Generally, cellulose shows some structural relations to beta-glucans, but due to the insolubility and functional dissimilarities, it is not considered as beta-glucan. The solubility of beta-glucan is enabled by the presence of beta-1,3 linkages (Buliga et al. 1986). Barley has the highest beta-glucan content (2.5–11.3%). Oats, rye, and wheat have 2.2–7.8%, 1.2–2%, and 0.4–1.4% of beta-glucans, respectively (Biliaderis and Izydorczyk 2007).

5.4 Solubilisation and Viscosity of Beta-Glucans

When ingested, beta-glucan becomes hydrated and swells. When digestion starts, beta-glucan is released and badly affects the digestion process. Depending on the nature of the feed the proportion of solubilised beta-glucan and the place of its solubilisation are varied in the gastrointestinal tract. In chicken, the beginning of solubilisation happens in the crop. The beta-glucan solubilisation increases with low pH, protein digestion, and vigorous physical action. In pigs, solubilisation ensues the digesta viscosity (Johansen et al. 1996). In pigs, the stomach digesta viscosity increases due to beta-glucan imbibition (Johansen et al. 1997). A study in humans also found that consumption of barley and oat beta-glucan rich food increases the viscosity of stomach digesta (Leuzinger et al. 2018). In the small intestine also, beta-glucan shows high viscosity. Animal's ability to digest more cell wall components and the changes in digesta pH increases beta-glucan aggregation and viscosity (Almirall et al. 1995). The enzymes produced by intestinal microbiota can hydrolyse beta-glucan and reduce viscosity. However, hydrolysis is dependent on the molecular weight of beta-glucan. In pigs, the viscosity reduction occurs through the transition of the duodenum and ileum. Sixty per cent soluble beta-glucan depolymerisation occurs in the duodenum and around 80% occurs in the ileum (Holtekjølen et al. 2014). In chickens, feed with high beta-glucan causes high ileal digesta viscosity. As a result of negligible microbial activity in the intestine, beta-glucan viscosity can only be reduced by the addition of exogenous enzymes (Kasprzak et al. 2012). Due to the less microbial enzyme activity in early weaned pigs, beta-glucan digestibility is very low. Hence, piglet feed also needs the addition of exogenous feed enzymes (Bach Knudsen et al. 1993).

5.5 Major Actions of Feed Enzymes in Animal Feeds

Exogenous feed enzymes have important roles in the animal feed industry. However, different feed enzymes (Table 5.1) have different modes of action. Some are as follows:

- Hydrolysis of specific bonds in the target substrates.
- Hydrolysis of anti-nutritional factors.
- Releasing of cell wall entrapped nutrients by disrupting endosperm integrity.
- The shift of nutrients to more competent digestion sites.
- Reducing endogenous secretions.
- Reducing protein loss (Cowieson and Ravindran 2007; Cowieson et al. 2009).
- Reducing the viscosity of intestinal digesta.
- Reducing the heaviness of the intestinal tract.
- Changing the morphology of the intestine (Jaroni et al. 1999; Wu et al. 2004).
- Improving host intestinal microbial profile (Bedford and Cowieson 2012; Apajalahti et al. 2004; Vahjen et al. 1998).
- Improving the digestion in animals with the immature digestive system.

The successive action of exogenous feed enzyme in the stomach and intestine depends on the source of enzymes, its specific activity, resistance proteolytic action, concentration and accessibility of substrates, and conditions in digestive tracts such as temperature, pH, moisture, etc. An aqueous environment in the stomach is necessary to initiate enzyme activity. When an animal ingests feed with an exogenous enzyme, the feed becomes quickly moistened for initiating the activity.

5.6 Role of Beta-Glucanases in Animal Feeds

Exogenous beta-glucanases are widely used as a feed enzyme in the animal feed industry to hydrolyse beta-glucan, a non-starch polysaccharide, in barley or oat containing diets. Due to the occurrence of beta-glucan in the cell wall of barley

Table 5.1 Different types of commercially available feed enzymes

Enzyme	Substrate	Target feed ingredient
Phytases	Phytic acid	All plant-derived ingredients
Beta-glucanases	Beta-glucan	Barley, oats and rye
Xylanases	Arabinoxylans	Wheat, barley, rye, triticale, fibrous plant material
Alpha-galactosidases	Oligosaccharides	Soybean meal, grain legumes
Proteases	Proteins	All plant protein sources
Amylase	Starch	Cereal grains, grain legumes
Lipases	Lipids	All plant lipid sources
Mannanases, cellulases, hemicellulases, pectinases	Cell wall matrix (fibre components)	Fibrous plant materials

and oat, the enzyme beta-glucanase is an apparent choice in animal feeds. Meanwhile, all these cereal grains also have arabinoxylans, the enzyme xylanase along with beta-glucanase enhances nutrient digestibility and growth statistics in animals (Owusu-Asiedu et al. 2010; Clarke et al. 2018; Munyaka et al. 2016). Both these enzymes increase nutrient digestibility and absorption by decreasing digesta viscosity. Mathlouthi et al. (2002a) confirmed the ability of xylanase and beta-glucanase in the viscosity reduction of non-starch polysaccharides in barley and oats. Both xylanase and beta-glucanase individually reduced the viscosity, but even more, the reduction was observed when used as a mixture of both enzymes. Dos Santos et al. also studied the synergic effect of beta-glucanase and xylanase in barley and oats-based diets. The addition of beta-glucanase as a supplement in animal feeds can diminish the negative effects of beta-glucans. Beta-glucans are the most critical constituents of barley and oat that increases digesta viscosity and constrains the nutrient's availability. Supplementation of beta-glucanase reduces digesta viscosity, improves essential microbiomes in the gastrointestinal tract and increases nutrient availability in monogastric animals (Raza et al. 2019; Fuente et al. 1995; Mathlouthi et al. 2002b; Józefiak et al. 2006; Yu et al. 1998). Most of the broiler diets that include barley, oats and wheat grains are supplemented with beta-glucanases. Beta-glucanase supplementation in poultry feed also reduces digesta viscosity, increases diet metabolizable energy and bird's growth performance (Perttilä et al. 2001; Józefiak et al. 2006). The practice of beta-glucanase preparation in the bird's diet results in a positive response and allows the increased use of cereals such as oats and barley.

High digesta viscosity and associated nutrient entrapment caused by beta-glucan rich feeds in animals can alleviate the effective use of exogenous beta-glucanase as a feed supplement. The potential advantages of exogenous beta-glucanase in animal diet to boost nutrient digestion and growth performance are explained well. However, physiological conditions in the stomach such as pH, protease activities, etc. may limit the retention time of enzymes in the gastrointestinal tract. Use of recombinant microbial beta-glucanase which shows better survival in wide ranges of pH and resistance to proteolytic cleavage is more advantageous to improve digestion. Intestinal digestibility of protein, starch and lipids are in between 85 and 90%, nevertheless, the addition of exogenous enzymes probably digests one-third of remaining undigested nutrients.

References

Almirall M, Francesch M, Perez-Vendrell AM, Brufau J, Esteve-Garcia E (1995) The differences in intestinal viscosity produced by barley and β-glucanase alter digesta enzyme activities and ileal nutrient digestibilities more in broiler chicks than in cocks. J Nutr 125(4):947–955

Annett CB, Viste JR, Chirino-Trejo M, Classen HL, Middleton DM, Simko E (2002) Necrotic enteritis: effect of barley, wheat and corn diets on proliferation of Clostridium perfringens type a. Avian Pathol 31(6):598–601

Apajalahti J, Kettunen A, Graham H (2004) Characteristics of the gastrointestinal microbial communities, with special reference to chickens. Worlds Poult Sci J 60(2):223–232

Bach Knudsen KE, Hansen I (1991) Gastrointestinal implications in pigs of wheat and oat fractions. 1. Digestibility and bulking properties of polysaccharides and other major constituents. Br J Nutr 65(2):217–232

Bach Knudsen KE, Jensen BB, Hansen I (1993) Digestion of polysaccharides and other major components in the small and large intestine of pigs fed on diets consisting of oat fractions rich in β-D glucan. Br J Nutr 70:531–556

Bedford MR, Cowieson AJ (2012) Exogenous enzymes and their effects on intestinal microbiology. Anim Feed Sci Technol 173(1–2):76–85

Bedford MR, Partridge GG (2001) Enzymes in farm animal nutrition: Cambridge. CABI Publishing, USA

Biliaderis CG, Izydorczyk MS (2007) Functional food carbohydrates: New York. CRC Press, NY, USA, p 21

Buliga GS, Brant DA, Fincher GB (1986) The sequence statistics and solution conformation of a barley (1,3-1,4)-β-d-glucan. Carbohydr Res 157:139–156

Campbell GL, Classen HL, Goldsmith KA (1983) Effect of fat retention on the rachitogenic effect of rye fed to broiler chicks. Poult Sci 62(11):2218–2223

Cho JH, Zhang ZF, Kim IH (2013) Effects of single or combined dietary supplementation of βglucan and kefir on growth performance, blood characteristics and meat quality in broilers. Br Poult Sci 54(2):216–221

Choct M, Kocher A, Waters DLE, Pettersson D, Ross G (2004) A comparison of three xylanases on the nutritive value of two wheats for broiler chickens. Br J Nutr 92(1):53–61

Clarke LC, Sweeney T, Curley E, Gath V, Duffy SK, Vigors S, Rajauria G, O'Doherty JV (2018) Effect of β-glucanase and β-xylanase enzyme supplemented barley diets on nutrient digestibility, growth performance and expression of intestinal nutrient transporter genes in finisher pigs. Anim Feed Sci Technol 238:98–110

Collett SR (2012) Nutrition and wet litter problems in poultry. Anim Feed Sci Technol 173(1–2): 65–75

Cowieson AJ, Bedford MR, Selle PH, Ravindran V (2009) Phytate and microbial phytase: implications for endogenous nitrogen losses and nutrient availability. Worlds Poult Sci J 65: 401–418

Cowieson AJ, Hruby M, Pierson EEM (2006) Evolving enzyme technology: impact on commercial poultry nutrition. Nutr Res Rev 19(1):90–103

Cowieson AJ, Ravindran V (2007) Effect of phytic acid and microbial phytase on the flow and amino acid composition of endogenous protein at the terminal ileum of growing broiler chickens. Br J Nutr 98(4):745–752

de Oliveira CAF, Vetvicka V, Zanuzzo FS (2019) β-Glucan successfully stimulated the immune system in different jawed vertebrate species. Comp Immunol Microbiol Infect Dis 62:1–6

Engstrom DF, Mathison GW, Goonewardene LA (1992) Effect of β-glucan, starch, and fibre content and steam vs. dry rolling of barley grain on its degradability and utilisation by steers. Anim Feed Sci Technol 37(1–2):33–46

Estrada A, Yun CH, Van Kessel A, Li B, Hauta S, Laarveld B (1997) Immunomodulatory activities of oat beta-glucan in vitro and in vivo. Microbiol Immunol 41(12):991–998

Ewing WR (1963) Poultry nutrition, 5th edn. Ray Ewing Co, South Pasadena, CA

Feighner SD, Dashkevicz MP (1988) Effect of dietary carbohydrates on bacterial cholyltaurine hydrolase in poultry intestinal homogenates. Appl Environ Microbiol 54(2):337–342

Fry RE, Allred JB, Jensen LS, James MG (1958) Effect of pearling barley and of different supplements to diets containing barley on chick growth and feed efficiency. McGin-nis. Poult Sci 37:281–288

Fuente JM, Pérezde Ayala P, Villamide MJ (1995) Effect of dietary enzyme on the metabolizable energy of diets with increasing levels of barley fed to broilers at different ages. Anim Feed Sci Technol 56(1–2):45–53

Guo Y, Ali RA, Qureshi MA (2003) The influence of β-glucan on immune responses in broiler chicks. Immunopharmacol Immunotoxicol 25(3):461–472

Hermans L, De Pelsmaeker S, Denaeghel S, Cox E, Favoreel HW, Devriendt B (2021) β-Glucan-induced IL-10 secretion by monocytes triggers porcine NK cell cytotoxicity. Front Immunol 12: 634402

Holtekjølen AK, Vhile SG, Sahlstrøm S, Knutsen SH, Uhlen AK, Åssveen M, Kjos NP (2014) Changes in relative molecular weight distribution of soluble barley beta-glucan during passage through the small intestine of pigs. Livest Sci 168:102–108

Hooda S, Metzler-Zebeli BU, Vasanthan T, Zijlstra RT (2011) Effects of viscosity and fermentability of dietary fibre on nutrient digestibility and digesta characteristics in ileal-cannulated grower pigs. Br J Nutr 106(5):664–674

Inborr J, Schmitz M, Ahrens F (1993) Effect of adding fibre and starch degrading enzymes to a barley/wheat based diet on performance and nutrient digestibility in different segments of the small intestine of early weaned pigs. Anim Feed Sci Technol 44(1–2):113–127

Jacob JP, Pescatore AJ (2014) Barley β-glucan in poultry diets. Ann Transl Med 2(2):20

Jaroni D, Scheideler SE, Beck MM, Wyatt C (1999) The effect of dietary wheat middlings and enzyme supplementation II: apparent nutrient digestibility, digestive tract size, gut viscosity, and gut morphology in two strains of leghorn hens. Poult Sci 78(12):1664–1674

Jensen LS, Fry RE, Allred JB, James MG (1957) Improvement in the nutritional value of barley for chicks by enzyme supplementation. McGin-nis Poult Sci 36:919–921

Johansen HN, Bach Knudsen KEB, Wood PJ, Fulcher RG (1997) Physico chemical properties and the degradation of oat bran polysaccharides in the gut of pigs. J Sci Food Agric 73(1):81–92

Johansen HN, Knudsen KEB, Sandström B, Skjøth F (1996) Effects of varying content of soluble dietary fibre from wheat flour and oat milling fractions on gastric emptying in pigs. Br J Nutr 75(3):339–351

Józefiak D, Rutkowski A, Jensen BB, Engberg RM (2006) The effect of β-glucanase supplementation of barley- and oat-based diets on growth performance and fermentation in broiler chicken gastrointestinal tract. Br Poult Sci 47(1):57–64

Kaldhusdal M, Hofshagen M (1992) Barley inclusion and avoparcin supplementation in broiler diets. 2. Clinical, pathological, and bacteriological findings in a mild form of necrotic enteritis. Poult Sci 71(7):1145–1153

Kasprzak MM, Lærke HN, Knudsen KEB (2012) Effects of isolated and complex dietary fiber matrices in breads on carbohydrate digestibility and physicochemical properties of ileal effluent from pigs. J Agric Food Chem 60(51):12469–12476

Koolman J, Roehm KH (2005) Color atlas of biochemistry, 2nd edn. Thieme Publishers, Stuttgart, Stuttgart, Germany

Langhout P (2000) New additives for broiler chickens. World Poult-Elsevier 16(3):22–27

Leuzinger S, Steingötter A, Nyström L (2018) Viscosity of cereal β-glucan in the gastrointestinal tract. Chimia 72(10):733–735

Mathlouthi N, Mallet S, Saulnier L, Quemener B, Larbier M (2002a) Effects of xylanase and β-glucanase addition on performance, nutrient digestibility, and physico-chemical conditions in the small intestine contents and caecal microflora of broiler chickens fed a wheat and barley-based diet. Anim Res 51(5):395–406

Mathlouthi N, Saulnier L, Quemener B, Larbier M (2002b) Xylanase, β-glucanase, and other side enzymatic activities have greater effects on the viscosity of several feedstuffs than xylanase and β-glucanase used alone or in combination. J Agric Food Chem 50(18):5121–5127

Mellor S (2000) Nutraceuticals-alternatives to antibiotics World Poult-Elsevier 16(2):30–33

Morales-López R, Auclair E, García F, Esteve-Garcia E, Brufau J (2009) Use of yeast cell walls; –1, 3. Poult Sci 88(3):601–607

Munyaka PM, Nandha NK, Kiarie E, Nyachoti CM, Khafipour E (2016) Impact of combined β-glucanase and xylanase enzymes on growth performance, nutrients utilization and gut microbiota in broiler chickens fed corn or wheat-based diets. Poult Sci 95(3):528–540

Murphy EA, Davis JM, Carmichael MD (2010) Immune modulating effects of beta-glucan. Curr Opin Clin Nutr Metab Care 13(6):656–661

Noss I, Doekes G, Thorne PS, Heederik DJJ, Wouters IM (2013) Comparison of the potency of a variety of β-glucans to induce cytokine production in human whole blood. Innate Immun 19(1): 10–19

Owusu-Asiedu A, Simmins PH, Brufau J, Lizardo R, Péron A (2010) Effect of xylanase and β-glucanase on growth performance and nutrient digestibility in piglets fed wheat-barley-based diets. Livest Sci 134(1–3):76–78

Perttilä S, Valaja J, Partanen K, Jalava T, Kiiskinen T, Palander S (2001) Effects of preservation method and β-glucanase supplementation on ileal amino acid digestibility and feeding value of barley for poultry. Br Poult Sci 42(2):218–229

Raza A, Bashir S, Tabassum R (2019) An update on carbohydrases: growth performance and intestinal health of poultry. Heliyon 5(4):e01437

Rodríguez ML, Rebolé A, Velasco S, Ortiz LT, Treviño J, Alzueta C (2012) Wheat- and barley-based diets with or without additives influence broiler chicken performance, nutrient digestibility and intestinal microflora. J Sci Food Agric 92(1):184–190

Sahasrabudhe NM, Tian L, Van den Berg M, Bruggeman G, Bruininx E, Schols HA, Faas MM, De Vos P (2016) Endo-glucanase digestion of oat β-glucan enhances Dectin-1 activation in human dendritic cells. J Funct Foods 21:104–112

Schwartz B, Vetvicka V (2021) Review: β-glucans as effective antibiotic alternatives in poultry. Molecules 26(12):3560

Selle PH, Ravindran V (2007) Microbial phytase in poultry nutrition. Anim Feed Sci Technol 135: 1–41

Synytsya A, Novak M (2014) Structural analysis of glucans. Ann Transl Med 2(2):17

Taylor PR, Brown GD, Reid DM, Willment JA, Martinez-Pomares L, Gordon S, Wong SY (2002) The β-glucan receptor, dectin-1, is predominantly expressed on the surface of cells of the monocyte/macrophage and neutrophil lineages. J Immunol 169(7):3876–3882

Teather RM, Wood PJ (1982) Use of Congo red-polysaccharide interactions in enumeration and characterization of cellulolytic bacteria from the bovine rumen. Appl Environ Microbiol 43(4): 777–780

Timbermont L, Haesebrouck F, Ducatelle R, Van Immerseel F (2011) Necrotic enteritis in broilers: an updated review on the pathogenesis. Avian Pathol 40(4):341–347

Vahjen W, Gläser K, Schäfer K, Simon O (1998) Influence of xylanase-supplemented feed on the development of selected bacterial groups in the intestinal tract of broiler chicks. J Agric Sci 130(4):489–500

Vetvicka V, Vetvickova J (2014) Natural immunomodulators and their stimulation of immune reaction: true or false? Anticancer Res 34(5):2275–2282

Viveros A, Brenes A, Pizarro M, Castaño M (1994) Effect of enzyme supplementation of a diet based on barley, and autoclave treatment, on apparent digestibility, growth performance and gut morphology of broilers. Anim Feed Sci Technol 48(3–4):237–251

Willingham HE, Fry RE, Leong KC, Jensen LS, McGinnis J (1961) Studies on stability of enzyme supplements to pelleting, long storage, and other treatments. Poult Sci 40(4):854–857

Wu YB, Ravindran V, Thomas DG, Birtles MJ, Hendriks WH (2004) Influence of phytase and xylanase, individually or in combination, on performance, apparent metabolisable energy, digestive tract measurements and gut morphology in broilers fed wheat-based diets containing adequate level of phosphorus. Br Poult Sci 45(1):76–84

Yegani M, Korver DR (2008) Factors affecting intestinal health in poultry. Poult Sci 87(10): 2052–2063

Yu B, Hsu JC, Chiou PWS (1998) Effects of β-glucanase supplementation of barley diets on growth performance of broilers. Anim Feed Sci Technol 70(4):353–361

Yun CH, Estrada A, Van Kessel A, Park BC, Laarveld B (2003) β-glucan, extracted from oat, enhances disease resistance against bacterial and parasitic infections. FEMS Immunol Med Microbiol 35(1):67–75

Chapter 6
Beta-Glucanase in Breweries

Lekshmi K. Edison, S. R. Reji, and N. S. Pradeep

Abstract The beta-glucan content from the endosperm cell wall of barley has a major impact on the quality of beer, especially during the filtration step, a major concern in the brewing industry. The high quantity and molecular weight of beta-glucans increase beer viscosity. Beta-glucan precipitates and forms gels and hazes. All these factors affect the yield of beer extraction. Grains with high beta-glucan content require lengthier germination during malting. Inactivation of endogenous beta-glucanase in cereal grains, during high-temperature processing steps in beer production considering the addition of exogenous commercial beta-glucanase in breweries as a substitute. It accelerates the overall process and preserves the quality of malt and beer. Understanding the physicochemical properties, hydrolysis and aggregation mechanisms of beta-glucan polymers helps to reduce the processing difficulties. This chapter summarizes the understanding of the barley beta-glucan, action and properties of endogenous barley beta-glucanases, application of exogenous microbial beta-glucanase during the malting and brewing processes.

Keywords Barley beta-glucan · Endogenous beta-glucanase · Exogenous beta-glucanase

6.1 Introduction

Beer is a common alcoholic beverage, mainly prepared using malted grain, hops, yeast and water. In the overall process, malt grain usually barley or wheat is milled and mixed with hot water known as mashing. During the mashing exercise, the

---

L. K. Edison
Rajiv Gandhi Centre for Biotechnology (RGCB), Thiruvananthapuram, Kerala, India

S. R. Reji
KSCSTE-Jawaharlal Nehru Tropical Botanic Garden and Research Institute, Thiruvananthapuram, Kerala, India

N. S. Pradeep (✉)
KSCSTE-Malabar Botanical Garden and Institute for Plant Sciences, Kozhikode, Kerala, India

N. S. Pradeep, L. K. Edison (eds.), *Microbial Beta Glucanases*, Interdisciplinary Biotechnological Advances, https://doi.org/10.1007/978-981-19-6466-4_6

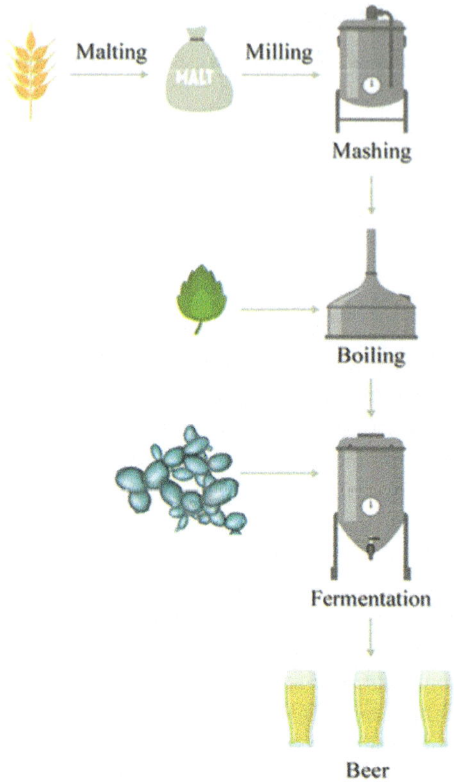

Fig. 6.1 Schematic representation of beer production

amylase enzymes hydrolyse the starch into fermentable carbohydrates. After mashing wort (sugar-rich lipid) is separated from the insoluble fraction known as brewer's spent grain, through the process called lautering. The wort mixed with hops boiled cooled and inoculated with brewer's yeast (Fig. 6.1). After fermentation, maturation, filtration and pasteurization, the finished product is packed into bottles (Dysvik et al. 2020). Malt is the major ingredient used in beer and whiskey production. It is a modified cereal grain product, getting after the germination of cereal grains using proper temperature, moisture and air. The process of making malt from cereal grain is known as malting. Barley (Hordeum vulgare) is the best malting grain; however, some other cereals are also used for the preparation of malt. The process of malting has 3 steps which include steeping, germination and kilning (Fig. 6.2). During malting, numerous changes occur in the cereal grain, include demolition of the endosperm cell walls, hydrolysis and controlled solubilization of proteins, production and activation of hydrolytic enzymes, etc. Complex biochemical and chemical reactions that happen during malting help the barley kernel to become more water-soluble (Briggs 1992, 1998; Briggs et al. 1981; Enari and Sopanen 1986).

Brewing industries require a huge amount of raw material for beer production. In 2016, one of the largest beer-producing countries Brazil produces around 43% of the

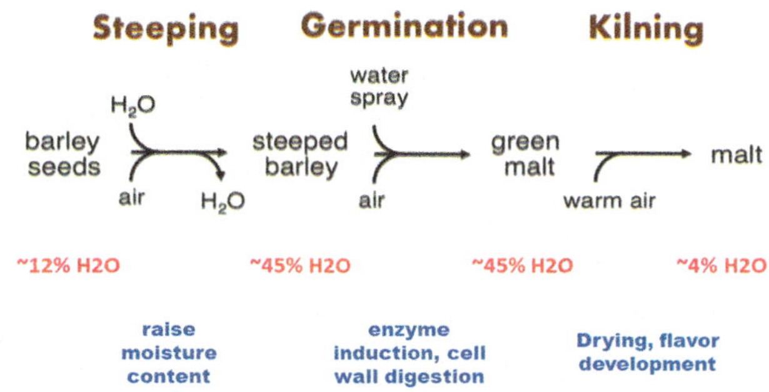

Fig. 6.2 The malting processes

malt for beer industries. However, due to the unavailability of barley and malt, many countries substitute and use rice, rye, oats, corn and sorghum (Kordialik-Bogacka et al. 2014; Poreda et al. 2014; Schnitzenbaumer et al. 2013; Hübner et al. 2010a, b; Depraetere et al. 2004). Cereals other than malt and barley decrease production cost, may adversely affect the physical-chemical properties of the beer and, in turn, reduces beer quality (Bogdan and Kordialik-Bogacka 2017). Intrinsic features of unmalted barley are facilitating its usage as an adjunct material in the brewing of beer. The temperature for the gelatinization is the same in both barley starch and malted barley starch.

Beta-glucanases (1,3-1,4-β-D-glucan 4-glucanohydrolase; 1,3-1,4-β-glucanase; EC 3.2.1.73) are the major enzymes formed during germination stage. It digests the cell wall components of barley grains. Hence, the replacement of malt by unmalted barley may affect the brewing process due to the high beta-glucan content and also due to the inadequate action of beta-glucanases (Farzaneh et al. 2017; Goode et al. 2005). Normal beta-glucan content in barley should be less than 178 mg. L^{-1}. The usage of unmalted barley increases higher beta-glucan content in beer, which causes to increase in the wort viscosity and turbidity, further reduces beer filtration rate (Wang et al. 2004; Kuntz and Bamforth 2007).

6.2 Barley Beta-Glucan [(1,3) (1,4)-β-D-Glucans]

Barley beta-glucans are the major component in barley endosperm cell walls (Fig. 6.3). It constitutes around 70% of total components (Palmer 1989). It is attached to the endosperm cell walls through protein-polysaccharide or phenol–ester linkages (Fincher 1975). Recommended beta-glucan levels in wort are 100–300 mg/L; however, the levels reported in many pieces of literature are 1152 mg/L (Jin 2002). It decreases the extract yield in breweries (Bamforth 1994; Edmunds et al. 1994; Evans et al. 1998; Haselmore et al. 1990; Henry 1986; Munck 1987). Reduced beta-glucan leads to the accumulation of unconverted protein and starch in the spent grain. These accumulated residuals reduce lautering and lead to loss of quality and quantity of final extract (Bathgate and Palmer 1973; Bamforth 1994; Palmer 1972). Beta-glucan increases apparent viscosities in the wort as well as

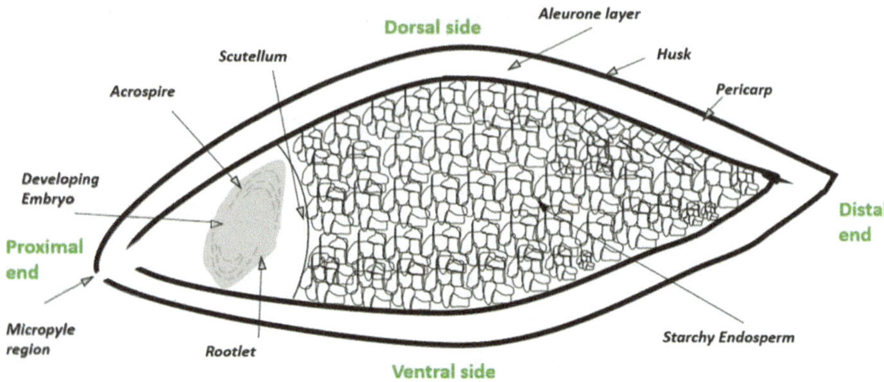

Fig. 6.3 Structure of barley grain (Neylon et al. 2020)

Table 6.1 Molecular weight distribution of barley beta-glucans during brewing process

Raw barley	150,000 to 1,937,000 KDa
Malt	800,000 to 1,220,000 KDa
Beer	1000 to 10,000,000 KDa

beer affects the filtration process (Charalambous 1981; Oonsivilai 2000; Oonsivilai et al. 1999; Patelakis 1999; Patelakis et al. 1999). Lastly, in the accumulation with proteins, pentosans and polyphenols beta-glucans form gel-like hazes in wort and beer (Igarashi and Amaha 1969; Whitear 1997). Thus, the malting quality is indicated by the content of barley β-glucans (Narziss et al. 1989; Haselmore et al. 1990). The molecular weight distribution of cereal beta-glucans during the brewing process is very important (Table 6.1). The molecular weight of barley beta-glucan varied in different stages of the brewing (Beer et al. 1997; Izawa et al. 1990; Wood et al. 1991a, b).

High molecular weight is mainly due to the molecular chain aggregation of beta-glucans. The precipitation of high molecular weight is very effortless after freezing and thawing. However, the precipitation of lower molecular weight beta-glucans is problematic during beer production. Beta-glucans with molecular weight less than 200,000 KDa remain in beer after the repeated steps of freezing and thawing (Izawa et al. 1990). The beta-glucan concentration in beer and the beer filtration rate are directly correlated (Nischwitz et al. 1999; Oonsivilai 2000). Continuous centrifugation, freezing and thawing and cross-flow filtration help the precipitation of beta-glucans but it clogs the membrane filters.

6.3 Endogenous Barley Beta-Glucanase

Beta-glucanase is an enzyme that hydrolyses the beta-glucan content in barley endosperm cell walls. They are classified into two types—exo acting beta-glucanase (exo beta-glucanase) and endo acting beta-glucanase (endo beta-glucanase) (Fig. 6.4). The activity level of beta-glucanase is based on the type of barley, development stage and malting conditions. Ungerminated barley grains detected inadequate hydrolysis of beta-glucan content due to insufficient enzyme activity caused high viscosity in beta-glucan extractions (Bathgate et al. 1974). During the first stage of beta-glucan degradation, the beta-glucan polymer in the endosperm cell wall solubilizes by the action of the beta-glucan solubilizing enzyme, beta-glucan solubilase (Bamforth and Quain 1989; Jin et al. 2004). Beta-glucan solubilase is one of the carboxypeptidases with esterolytic activity. It was characterized as heat-stable (Bamforth et al. 1997; Yin and MacGregor 1989) and the activity was detected in both barley and malt (Palmer 1975; Yin and MacGregor 1988, 1989). Later,

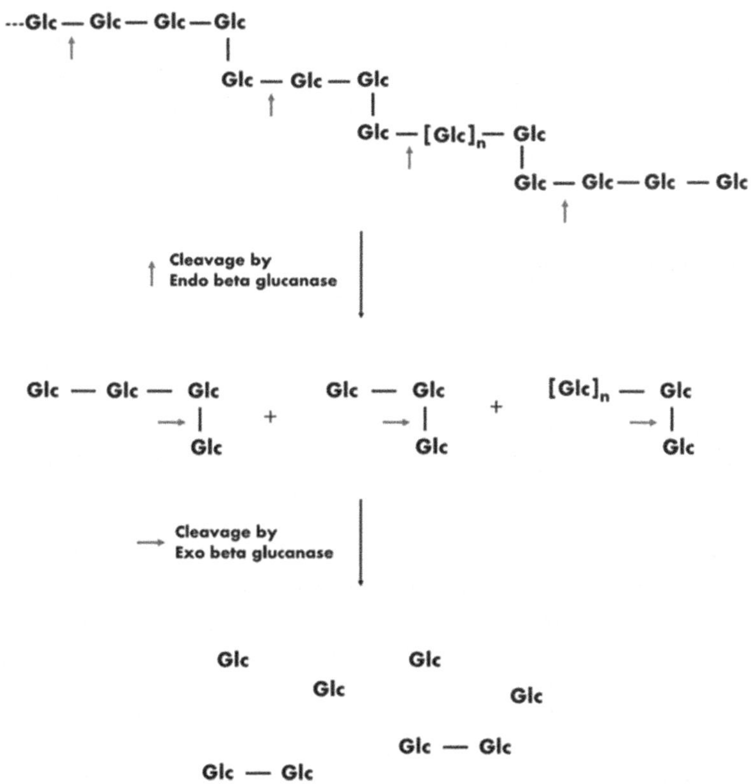

Fig. 6.4 Action of exo acting beta-glucanase and endo acting beta-glucanase on cereal beta-glucan. Endo acting beta-glucanase randomly cleaves internal glycosidic bonds and exo acting beta-glucanase cleaves the non-reducing ends of the beta-glucan chain and releases glucose units

Table 6.2 Properties of the isoenzymes of barley endo beta-glucanase

Property	Isoenzyme EI	Isoenzyme EII
Amino acids	306	306
Approximate molecular weight (Da)	30,000	32,000
Isoelectric point	8.5	10.6
pH optimum	4.7	4.7
Carbohydrate content (%)	Trace	~4
Thermal stability	Low	Greater
Secretion site	Scutellum, and aleurone layer	Aleurone layer

beta-glucan solubilase has been recognized as endo (1,3)-beta-glucanase and endo (1,4)-beta-glucanase (Bathgate et al. 1974; Yin and MacGregor 1988). Beta-glucanase activity increases during germination, however, slightly decreases at the time of kilning (Hübner et al. 2010a, b). The enzyme activity was detected at 62 °C for 30 minutes (Sim and Berry 1996).

Initial degradation of beta-glucan during malting and mashing is done by the enzyme endo beta-glucanases. It reduces beta-glucan molecular size. Endo beta-glucanase has two enzymatic activities, such as laminarinase (EC 3.2.1.6) and lichenase (EC 3.2.1.73) activities (Poutanen 1997). Lichenase is commercially available, *Bacillus subtilis* derived beta-glucanase. It is also known as beta-(1,3)(1,4)-D-glucan 4-glucanohydrolase or endo beta-(1,3)(1,4)-glucanase, which precisely hydrolyses beta-1,4-linkages of the O-3-substituted glucose of barley beta-glucans (Mcclear and Glennie-Holmes 1985; Müller et al. 1998; Wood et al. 1991a, b).

Laminarinase cleaves both beta (1,3) and beta (1,4) glucosidic bonds in barley β-glucans. Synthesis of the considerable amount of endo beta-glucanase in raw barley takes place by de novo pathway which occurs during the early stages of grain germination and stimulates by gibberellic acid (GA) (Jin and Liu 1994a, b; Manners et al. 1971; Palmer 1973; Yomo and Iinuma 1964). The activity of endo beta-glucanase is slightly increased during steeping when the moisture content increased up to 46%. After three days of germination, activity reaches a maximum (Lee et al. 1998). Around 62% of the endo-β-glucanase activity survives during kilning. Though, high temperatures drying or kilning may cause sudden enzyme inactivation (Bourne et al. 1976). Endo beta-glucanases in malt are highly sensitive to heat (Bamforth 1985; Bourne et al. 1976; Home et al. 1993; Loi et al. 1987). Because of the high heat in the malt solution, the enzyme activity is severe. However, in thicker mashes, enzymes possess high thermostability due to a shielding effect of reduced glutathione. In addition to another reducing agent, DL-Dithiothreitol also gives protection on the active site residue (Met7) during the storage (Müller et al. 1998; Brunswick et al. 1988).

The actual hydrolysis of barley beta-glucan happens by mixed action two isoenzymes of endo beta-glucanase—EI and EII, which are produced by diverse tissues and the production is under distinct genetic controls (Stuart et al. 1986). Both isoenzymes are slightly different in their properties (Table 6.2) (Izydorczyk and

Table 6.3 Properties of the isoenzymes of barley exo beta-glucanase

Property	ExoI	ExoII
Amino acids	627	645
Approximate molecular weight (Da)	69,000	71,000
Isoelectric point	7.8	8.0
pH optimum	5.25	5.25
Thermal stability	Low	Low
Secretion site	Scutellum	Scutellum

Edney 2003). Studies pointed out that there is no ostensible connection with the malt endo beta-glucanase enzyme level and activity with the degree of beta-glucan hydrolysis (Nischwitz et al. 1999). Even if the activity of in vivo beta-glucanase is lower, degradation of beta-glucan occurred during the early days of germination (Palmer et al. 1985).

Another enzyme, exo beta-glucanase is detected in barley during germination. But its activity showed relatively lower during the early stages of germination and increased 3 to 4 times during late germination (Fincher and Stone 1993). It acts on the non-reducing ends of the beta-glucan molecule and releases glucose units. It is involved in (1,3), (1,4) beta-glucan hydrolysis and is named beta-glucan glucohydrolases and beta-glucosidase. Exo beta-glucanase has two isoenzymes—ExoI and ExoII (Hrmova et al. 1996). Properties of two isoenzymes are listed in Table 6.3. Hence the thermostability is very low, the activity loss of exo beta-glucanase is more severe during kilning. During the activity on barley beta-glucan, the enzyme exo beta-glucanase releases only one glucose unit at every cleavage. So, the impact of the breakdown of the beta-glucan polymer in the reduction of wort viscosity is very less in contrast to endo beta-glucanases.

6.4 Beta-Glucan Degradation During Malting

The factors affecting malting are hydration of the endosperm, fast and unvarying germination and an increase in GA response. All these factors confirm the production, distribution and activity of endosperm hydrolysing enzymes. The rate of beta-glucan degradation is affected by the size and denseness of the endosperm wall, size and structure of beta-glucan and the number of lytic enzymes (Oksanen et al. 1985; Aastrup and Munck 1985; Narziss et al. 1989; Palmer et al. 1985). The removal of the arabinoxylan layer from the aleurone cell wall is essential for the distribution and migration of endo beta-glucanase to the beta-glucan layer (Smart et al. 1993; Autio et al. 1996). Early release of beta-glucanase was observed in steeped barley (Manners and Marshall 1969). Malt quality is a major factor defining the beta-glucan levels in the wort. During malting, the quality of raw barley affects beta-glucan degradation. Hence, the use of low beta-glucan barley breed can overcome the problems related to beta-glucan degradation during brewing (Schildbach 1998).

In cereal grains, beta-glucan degradation is very slow during the steeping stage. Initial beta-glucan degradation is taking place during the germination stage of malting. During early days of germination (first 3 days) the degradation is very fast and steady. During the late stages of germination and also in killing the beta-glucan degradation declines (Bamforth 1985). Prolonged steeping time (30 hrs instead of 18 hrs) with high steep water temperature (22 °C instead of 16 °C) help to diminish the wort beta-glucans level. Higher temperature helps to rise the endo beta-glucanase level during germination. But higher temperature increases protein degradation. Proteins are essential for the foaming of beer. It also lowers the level of malt extract (Nischwitz et al. 1999). Beta-glucan hydrolysis can be accelerated by germinating the grains with 45–48% of moisture content. The gibberellic acid application reduces the time of seed germination and accelerates the beta-glucan hydrolysis (Smart et al. 1993; Bourne and Wheeler 1984; Anderson et al. 1989). According to Autio et al. (1996), the purified endo beta-glucanase isoenzymes, EI and EII, have shown the same pattern of hydrolysis when incubating with barley grain sections. Though, the cell wall degradation was not uniform. The enzymes initially attacked the endosperm cell wall.

Kilning with the greater grain bed depths, especially a double-floor kiln, the hydrolysis of beta-glucans remains. The lower temperature and higher moisture in the top layer of skin favour beta-glucan degradation (Anderson et al. 1989). The beta-glucan content and beta-glucanase activity from the bottom, middle and top grain bed is not static. The highest beta-glucan content and lowest beta-glucanase activity were observed in the bottom grain bed. The decreased beta-glucan level in the upper layer is possibly due to the prolonged treatment with 30–35 °C of temperature and around 45% of moisture content (Piglas et al. 1988).

6.5 Beta-Glucan Degradation During Mashing

Mashing temperature is one of the critical factors affecting beta-glucan degradation. Malting temperature is directly correlated with beta-glucan precipitation potential. Mashing at 65 °C temperature inactivates endogenous beta-glucanases in cereal grains. The partially degraded beta-glucan forms polymers, which precipitates and forms gels during freezing and thawing cycles of the brewing process. It increases wort viscosity because 70% of beta-glucan is dissolved in sweet wort, after cooling around 30% of beta-glucan in wort precipitates (Letters et al. 1985; Izydorczyk et al. 1998a, b). Multiple mashing with a lower mashing temperature of around 48 °C with a long retention time stabilizes the beta-glucanases activity (Narziss et al. 1990; Kettunen et al. 1996; Home et al. 1993). High mashing temperature also causes starch gelatinization. Hence lower temperature mashing with longer holding periods helps to hydrolyse sufficient beta-glucan content before saccharification (Bamforth and Quain 1989).

Lowering the beta-glucan content to less than 200,200 mg/L requires sufficient production and retention of beta-glucanase activity in the malt is essential

(Annemuller et al. 1996). Endogenous endo beta-glucanase enzyme reduces maximum wort viscosity at 40 to 45 °C temperatures with pH 4.5 to 4.8. While, exo beta-glucanase shows optimal activity at 40 °C and pH 4.5 and contributes a diminutive reduction in wort viscosity (Narziss 1973, 1993). During mashing, the glucanolytic enzymes are slowly inactivated at temperature 40 °C and quickly at 50 °C. Very low activity was detected at 55 °C and no activity was at higher temperatures (Home et al. 1993; Loi et al. 1987). Proper standardization of mash thickness and agitation speed can also help to reduce the wort beta-glucan level. The addition of commercially available thermostable beta-glucanase is another to overcome the problems related to high beta-glucan content during mashing.

6.6 Microbial Beta-Glucanases in Breweries

In breweries, beta-glucan related problems are primarily connected with when barley is used as a modified malt. During malting, under modified barley leaves more gummy spent grains, causes the formation of gels and hazes and lowers filtration rates (Wainwright 1999). Under modification is mainly due to damage in cereal grain, dormant stage, improper absorption of gibberellic acid, uneven grains from different corns, variation in oxygen supply and temperature during germination and steeping, etc. (Jin and Liu 1994a, b; Hyde and Brookes 1978). Brewing industries are constantly looking for solutions to upgrade the malting process to make them cost-effective (Müller and Methner 2015). However, changes in steeping, sprouting or desiccating may affect the beer quality. The addition of exogenous enzymes is the only solution to improve the quality of malt as well as beer. Beta-glucanase preparations from the microbial source are promising to decrease the beta-glucan content in barley (Goode et al. 2005; Hattingh et al. 2014). Its usage prevents the formation of beta-glucan gels and hazes, improves filtration rate and increases the yield of extract (Wainwright 1999; Pool and Pollock 1971). Studies showed that both barley and *Bacillus subtilis* endo beta-glucanases have similar substrate specificity and stereochemistry (Müller et al. 1998; Malet et al. 1993).

The addition of heat-stable microbial beta-glucanases can be done during any stages of the brewing process such as malting, mashing or fermentation, where it reduces high molecular weight beta-glucan content in the wort, leading to better filtration rate and improved beer production (Grujić 1998; Izydorczyk et al. 1998a, b; Kettunen et al. 1996; Todo et al. 1989; McKechnie and O'Sullivan 1995; Narziss et al. 1990; Sudarmana et al. 1996). Different commercial beta-glucanase preparation has varying efficiency in reducing beta-glucan levels (Ducroo and Frelon 1989). According to Brazil et al. (2019), 50 mg.kg^{-1} of commercial beta-glucanase from AB Enzymes, Darmstadt, Germany is enough to produce a good quality malt. A 1,3-1,4-β-D-glucan 4-glucan hydrolase produced by *rhizopus microsporus* var. *microsporus* showed activity optimum at pH 4–5 and temperature 50–60 °C. Kinetic study and the ability to reduce mash viscosity and filtration time indicated its efficient use in the brewing industry (Celestino et al. 2006). Addition of 20 U/ml of endo beta (1,3) (1,4)-glucanase from thermoacidophilic *Alicyclobacillus* sp. A4 more effectively reduced viscosity and increased filtration rate suggesting the efficient use in breweries (Bai et al. 2010). The addition of thermostable beta-glucanases is preferred during mashing because its stability at a high temperature helps to increase the yield as well as it can be inactivated by the boiling of wort.

References

Aastrup S, Munck L (1985) A β-glucan mutant in barley with thin cell walls. In: Hill RD, Munck L (eds) New approaches to research on cereal carbohydrates. Elsevier Science Ltd., Amsterdam, Netherlands, pp 291–296

Anderson IW, Dickenson CJ, Anderson RG (1989) β-Glucan malting. In: Proceedings of the congress Eur brew conv Zurich 22. IRL Press at Oxford University Press, Oxford, UK, pp 213–220

Annemuller G, Bauch T, Nagel R, Böhm W (1996) Endo-β-glucanase levels–A measure for the "cytolytic power" of malt? Brauwelt Int 14:312–317

Autio K, Mannonen L, Pietilä K, Koskinen M, Suka-aho M, Linko M, Morgan A (1996) Incubation of barley kernel sections with purified cell wall degrading enzymes. J Inst Brew 102(6):427–432

Bai Y, Wang J, Zhang Z, Shi P, Luo H, Huang H, Luo C, Yao B (2010) A novel family 9 β-1,3 (4). Appl Microbiol Biotechnol 87(1):251–259

Bamforth CW (1994) β-Glucan and β-glucanases in malting and brewing: practical aspects. Brew Dig 69(5):12–16, 21

Bamforth CW (1985) Biochemical approaches to beer quality. J Inst Brew 91(3):154–160

Bamforth CW, Moore J, McKillop D, Williamson G, Kroon PA (1997) Enzymes from barley which solubilize β-glucan. In: Proceedings of the congress Eur brew conv Maastricht 26, 75–82. IRL Press at Oxford University Press: Oxford, U.K.

Bamforth CW, Quain DE (1989) Enzymes in brewing and distilling. In: Palmer GH (ed) Cereal science and technology. Aberdeen University Press, Aberdeen, U.K, pp 326–366

Bathgate GN, Palmer GH (1973) The in vivo and in vitro degradation of barley and malt starch granules. J Inst Brew 79(5):402–406

Bathgate GN, Palmer GH, Wilson G (1974) The action of endo-β-glucanases on barley and malt β-glucans. J Inst Brew 80(3):278–285

Beer MU, Wood PJ, Weisz J (1997) Molecular weight distribution and (1→3)(1→4)-β-D-glucan content of consecutive extracts of various oat and barley cultivars. Cereal Chem 74(4):476–480

Bogdan P, Kordialik-Bogacka E (2017) Alternatives do malt in brewing. Trends Food Sci Technol 65:1–9

Bourne DT, Jones M, Pierce JS (1976) Beta glucan and beta glucanase in malting and brewing. Tech Q Master Brew Assoc Am 13:3–7

Bourne DT, Wheeler RE (1984) Environmental and varietal differences in total β-glucan contents of barley and the effectiveness of its breakdown under different malting conditions. J Inst Brew 90(5):306–310

Brazil C, de Oliveira DF, Duarte RA, Galo JM, Lucchetta L, Santos EC, Hashimoto EH (2019) β-Glucanase addition in brewing malt produced by reduced time of germination. Braz Arch Biol Technol 62:e19180315

Briggs DE (1992) Barley germination: Biochemical changes and hormonal control. In: Shewry PR (ed) Barley: genetics, biochemistry, molecular biology and biotechnology. CA B International, Wallingford, UK, pp 369–401

Briggs DE (1998) Malts and malting. Blackie Academic & Professional, London

Briggs DE, Hough JS, Stevens T, Young TW (1981) Malting and brewing science 1. In: Malt and sweet wort, 2nd edn. Chapman & Hall, London

Brunswick P, Manners DJ, Stark JR (1988) Degradation of isolated barley endosperm cell walls by purified endo-(1→3)(1→4)-β-D-glucanases and malt extracts. J Cereal Sci 7(2):153–168

Celestino KRS, Cunha RB, Felix CR (2006) Characterization of a β-glucanase produced by *Rhizopus microsporus* var. *microsporus*, and its potential for application in the brewing industry. BMC Biochem 7:7:23

Charalambous G (1981) Involatile constituents of beer. In: Pollock JRA (ed) Brewing science, vol 2. Academic Press, Inc, New York, pp 167–254

Depraetere SA, Delvaux F, Coghe S, Delvaux FR (2004) Wheat variety and barley malt properties: influence on haze intensity and foam stability of wheat beer. J Inst Brew 110(3):200–206

Ducroo P, Frelon PG (1989) Improvement of beer production by the use of β-glucanase-pentosanase from Disporotrichum dimorphosporum. In: Proceedings of the congress Eur brew conv Zurich 22, 445–452. IRL Press at Oxford University Press: Oxford, U.K.

Dysvik A, La Rosa SL, De Rouck G, Rukke EO, Westereng B, Wicklund T (2020) Microbial dynamics in traditional and modern sour beer production. Appl Environ Microbiol 86(14): e00566–e00520

Edmunds MD, Allan GR, Fincher GB, Stuart IM (1994) The quantification of (1→3, 1→4)-β-glucanase isoenzymes during malting using an enzyme-linked immunosorbent assay. In: Proceedings of the Conv. Inst. Brew, Pacific sect A. 23rd, Sydney, London, U.K: Institute of Brewing. p 79-83

Enari T-M, Sopanen T (1986) Mobilisation of endospermal reserves during the germination of barley. J Inst Brew 92(1):25–31

Evans E, Stenholm K, Vilpola A, Home S, Hughes G (1998) Pilot scale evaluation of mash filter performance of malts with variable quality. Tech Q Master Brew Assoc Am 35:189–195

Farzaneh V, Ghodsvali A, Bakhshabadi H, Zare Z, Carvalho IS (2017) The impact of germination time on the some selected parameters through malting process. Int J Biol Macromol 94 (A):663–668

Fincher GB (1975) Morphology and chemical composition of barley endosperm cell walls. J Inst Brew 81(2):116–122

Fincher GB, Stone BA (1993) Physiology and biochemistry of germination in barley. In: MacGregor AW, Bhatty RS (eds) Barley: chemistry and technology. American Association of Cereal Chemists St. Paul, MN, pp 247–295

Goode DL, Wijngaard HH, Arendt EK (2005) Mashing with unmalted barley–impact of malted barley and commercial enzyme (Bacillus spp.) additions. Tech Q Master Brew Assoc Am 42: 184–198

Grujić O (1998) Application of a commercial enzyme preparation in the barley malting process. J Inst Brew 104(5):249–253

Haselmore RM, Tunnicliffe CG, Peipi MG, Slack CR (1990) Using wort β-glucan levels to aid selection in a barley breeding programme. In: Proceedings of the Conv: Aust: Inst. Brew NZ Sect, ed., 21st, Auckland, pp, 79–82: London, U.K: Institute of Brewing

Hattingh M, Alexander A, Meijering I, van Reenen CA, Dicks LMT (2014) Malting of barley with combinations of lactobacillus plantarum, aspergillus Niger, Trichoderma reesei, Rhizopus Oligosporus and Geotrichum candidum to enhance malt quality. Int J Food Microbiol 173: 36–40

Henry RJ (1986) Genetic and environmental variation in the pentosan and β-glucan contents of barley, and their relation to malting quality. J Cereal Sci 4(3):269–277

Home S, Pietilä K, Sjöholm K (1993) Control of glucanolysis in mashing. J Am Soc Brew Chem 51(3):108–113

Hrmova M, Harvey AJ, Wang J, Shirley NJ, Jones GP, Stone BA, Høj PB, Fincher GB (1996) Barley beta-D-glucan exohydrolases with beta-D-glucosidase activity. Purification, characterization, and determination of primary structure from a cDNA clone. J Biol Chem 271(9): 5277–5286

Hübner F, O'Neil T, Cashman KD, Arendt EK (2010a) The influence of germination conditions on beta-glucan, dietary fibre and phytate during the germination of oats and barley. Eur Food Res Technol 231(1):27–35

Hübner F, Schehl BD, Gebruers K, Courtin CM, Delcour JA, Arendt EK (2010b) Influence of germination time and temperature on the properties of rye malt and rye malt based worts. J Cereal Sci 52(1):72–79

Hyde WR, Brookes PA (1978) Malt quality in relation to beer quality. J Inst Brew 84(3):167–174

Igarashi H, Amaha M (1969) Frozen beer precipitates. II. Chemical structure of the β-glucan isolated from the precipitates. J Inst Brew 75(3):292–299

Izawa M, Kano Y, Kamimura M (1990) β-Glucan responsible for haze formation in beer. Curr Dev Malting Brew Distill 3:427–430

Izydorczyk MS, Edney MJ (2003) MALT. Chemistry of malting. In: Caballero B (ed) Encyclopedia of food science and nutrition, 2nd edn. Academic Press, pp 3677–3685

Izydorczyk MS, Macri LJ, MacGregor AW (1998a) Structure and physiochemical properties of barley non-starch polysaccharides–I. Water-extractable β-glucans and arabinoxylans Carbohydr Polym 35(3–4):249–258

Izydorczyk MS, Macri LJ, MacGregor AW (1998b) Structure and physiochemical properties of barley non-starch polysaccharides–II. Alkali-extractable β-glucans and arabinoxylans Carbohydr Polym 35(3–4):259–269

Jin Y-L (2002) Effect of β-glucan and environmental factors on the physical and chemical properties of wort and beer [Ph.D. thesis]: Halifax, NS, Canada: Dalhousie University

Jin Y-L, Liu Z-Q (1994a) Oxygen uptake and aeration during steeping and germination. Damei Kexue (Barley Sci) 46(3):41–44

Jin Y-L, Liu Z-Q (1994b) Studies on malt endo-β-glucanase Damei Kexue (Barley Sci) 46(1):36–38

Jin YL, Speers RA, Paulson AT, Stewart RJ (2004) Barley β-glucans and their degradation during malting and brewing. Master Brew Assoc Am-Tech Q 41(3):231–240

Kettunen A, Hämäläinen JJ, Stenholm K, Pietilä K (1996) A model for the prediction of β-glucan concentration during mashing. J Food Eng 29(2):185–200

Kordialik-Bogacka E, Bogdan P, Diowksz A (2014) Malted and unmalted oats in brewing. J Inst Brew 130:n/a-n/a

Kuntz RJ, Bamforth CW (2007) Time course for the development of enzymes in barley. J Inst Brew 113(2):196–205

Lee R, Allan G, Greig R, Dunn C, Stuart M (1998) The effect of (1→3, 1→4)-β-glucan and arabinoxylan on malt quality and brewhouse performance. In: Proceedings of the Conv. Inst. Brew, Pacific Sect A. 25th, Perth, 33–37: London, U.K: Institute of Brewing

Letters R, Byrne H, Doherty M (1985) The complexity of beer β-glucans. In: Proceedings of the congress Eur brew conv Helsinki 20, 395-402. IRL Press at Oxford University Press: Oxford, U.K.

Loi L, Barton PA, Fincher GB (1987) Survival of barley (1→3, 1→4)-β-glucanase isozymes during kilning and mashing. J Cereal Sci 5(1):45–50

Malet CJ-B, Bernabé M, Brosa C, Planas A (1993) Sterochemical course and structure of the products of the enzymic action endo-1,3-1,4-β-D-glucan 4-glucanohydrolase from bacillus licheniformis. Biochem J 296:753–758

Manners DJ, Marshall JJ (1969) Studies on carbohydrate metabolizing enzymes, XXII. The β-glucanase system of malted barley. J Inst Brew 75(6):549–561

Manners DJ, Palmer GH, Wilson G, Yellowlees D (1971) Effect of gibberellic acid on the development of some cereal carbohydrases. Biochem J 125(2):30P–31P

Mcclear BV, Glennie-Holmes M (1985) Enzymic quantification of (1→3)(1→4)-β-glucan in barley and malt. J Inst Brew 91(5):285–295

McKechnie MT, O'Sullivan P (1995) Filter fouling with molecular species: stirring up the evidence. BRFI Q. April:29-31

Müller C, Methner FJ (2015) An accelerated malting procedure- influences on malt quality and cost saving by reduced energy consumption and malting losses. J Inst Brew 121(2):181–192

Müller JJ, Thomsen KK, Heinemann U (1998) Crystal structure of barley 1,3-1,4-β-glucanase at 2.0-Å resolution and comparison with bacillus 1,3-1,4-β-glucanase. J Biol Chem 273(6): 3438–3446

Munck L. (1987). Quality assessment of brewing barley varieties. In: Proceedings of the congress Eur brew conv Madrid 21, 145–156. IRL Press at Oxford University Press: Oxford, U.K.

Narziss L (1973) Zusammenhänge zwischen Gerstenbeschaffenheit, Malz und Bierqualitaet [Correlation between quality of barley, malt and beer]. Brauwelt 113:3–8

Narziss L (1993) Beta-glucan and filterability Brauwelt Int 11:435–442

Narziss L, Reicheneder E, Edney MJ (1989) Importance of beta-glucan size and concentration in malting. Monatsschr Brauwiss 42:430–437

Narziss L, Reicheneder E, Edney MJ (1990) The control of β-glucan in the brewery. Monatsschr Brauwiss 43:66–76

Neylon E, Arendt EK, Lynch KM, Zannini E, Bazzoli P, Monin T, Sahin AW (2020) Rootlets, a malting by-product with great potential. Fermentation 6(4):117

Nischwitz R, Cole NW, MacLeod L (1999) Malting for brewhouse performance. J Inst Brew 105(4):219–228

Oksanen J, Ahvenainen J, Home S (1985) Microbial cellulase for improving filterability of wort and beer. In: Proceedings of the congress Eur brew conv Helsinki 20, 419–425. IRL Press at Oxford University Press: Oxford, U.K.

Oonsivilai R (2000) The effect of beta-glucan polymers on the rheological and filtration properties of wort [M.Sc. thesis]: Halifax, NS, Canada: Dalhousie University

Oonsivilai R, Patelakis S, Speers RA, Paulson AT (1999) Rheological and filtration properties of β-glucan polymers in the brewing process. In: Food, health and science. Annu conference Canada Institucionais Food Sci. Technol., 41st. Toronto, ON, Canada: Canadian Institute of Food Science and Technology (CIFST). p 40

Palmer GH (1972) Morphology of starch granules in cereal grains and malts. J Inst Brew 78(4): 326–332

Palmer GH (1973) Relationships between levels of gibberellic acid and the production and action of carbohydrate of barley. J Inst Brew 79(6):513–518

Palmer GH (1975) Influence of endosperm structure on extract development. Proc Am Soc Brew Chem 33:174–180

Palmer GH (1989) Cereals in malting and brewing. In: Cereal science and technology, 61-242 GH Palmer, ed.: Aberdeen, U.K: Aberdeen University Press

Palmer GH, Gernah DI, McKernan G, Nimmo DH, Laycock G (1985) Influence of enzyme distribution on endosperm breakdown (modification) during malting. J Am Soc Brew Chem 43(1):17–28

Patelakis S (1999) The effect of beta-glucan polymers on the rheological and filtration properties of beer [M.Sc. thesis]. DalTech: Halifax, NS, Canada: Dalhousie University

Patelakis S, Speers RA, Paulson A, Stewart (1999) Effect of β-glucan on the rheological and filtration properties of beer [Abstr.] Am Soc Brew Chem Newsl R J 59(2):16

Piglas J, Coates KD, Armitt JDG (1988) Malt inhomogeneity induced by kilning. In: Proceedings of the Conv: Aust: Inst. Brew NZ Sect, ed., 20th, Brisbane, pp, 81–85: London, U.K: Institute of Brewing

Pool AA, Pollock RA (1971) Wort viscosity control. British patent patent specification 1243368

Poreda A, Czarnik A, Zdaniewicz M, Jakubowski M, Antkiewicz P (2014) Corn grist adjunct–application an influence on the brewing process and beer quality. J Inst Brew 120(1):77–81

Poutanen K (1997) Enzymes: an important tool in the improvement of the quality of cereal foods. Trends Food Sci Technol 8(9):300–306

Schildbach R (1998) The contribution of malting barley cultivars to rationalization of malting. Brauwelt 138(178–181):184–185

Schnitzenbaumer B, Karl CA, Arendt EKA (2013) Comparison of white Nigerian and red Italian sorghum (Sorghum bicolor) as brewing adjuncts based on optimized enzyme additions. J Am Soc Brew Chem 71(4):248–257

Sim GB, Berry DR (1996) Malted barley enzyme activity under optimum and process conditions from the scotch malt whisky industry. Enzym Microb Technol 19(1):26–31

Smart JG, Lukes BK, Tie E, Ford AT (1993) Relationships among wort β-glucan, malting conditions, and malt analysis. J Am Soc Brew Chem 51(3):88–93

Stuart IM, Loi L, Fincher GB (1986) Development of (1→3, 1→4)-β-D-glucan endohydrolase isozymes in isolated scutella and aleurone layers of barley (Hordeum vulgare). Plant Physiol 80(2):310–314

Sudarmana DL, Goldsmith MR, Hinh AH, Pecar MA, Hawthorne DB, Kavanagh TE (1996) Microfiltration studies with a modified membrane filterability procedure. Tech Q Master Brew Assoc Am 33:63–72

Todo V, Carbonell JV, Sendra JM (1989) Kinetics of β-glucan degradation in wort by exogenous β-glucanases treatment. J Inst Brew 95(6):419–422

Wainwright T (1999) Enzymes in brewing. Ferment 12:30–34

Wang J, Zhang G, Chen J, Wu F (2004) The changes of β-glucan content and β-glucanase activity in barley before and after malting and their relationships to malt qualities. Food Chem 86(2): 223–228

Whitear T (1997) Dealing with persistent hazes in freshly filtered beer. Ferment 10:186–190

Wood PJ, Weisz J, Fedec P (1991a) Potential for β-glucan enrichment in brans derived from oat (Avena sativa L.) cultivars of different (1→3),(1→4)-β-D-glucan concentrations. Cereal Chem 68:48–51

Wood PJ, Weisz J, Mahn W (1991b) Molecular characterization of cereal β-glucans. II. Size-exclusion chromatography for comparison of molecular weight. Cereal Chem 68:530–536

Yin XS, MacGregor AW (1988) An approach to the identification of a β-glucan solubilase from barley. J Inst Brew 94(5):327–330

Yin XS, MacGregor AW (1989) Substrate specificity and nature of action of barley β-glucan solubilase. J Inst Brew 95(2):105–109

Yomo H, Iinuma H (1964) The enzymes of the aleurone layer of barley endosperm. Proc Am Soc Brew Chem 22(1):97–102

Chapter 7
Applications of Microbial Beta-Glucanase in Crop Improvement Under Biotic and Abiotic Stress

Yachana Jha

Abstract Plant-associated microbes have evolved in such a way that now such associations play important role in a healthy ecosystem. Plant-associated microbes facilitate the host plant with different types of important metabolites for its proper growth under normal as well as under stress. The microbial β-glucanases produced by the plant-associated bacteria have also a versatile role in plant growth physiology as well as help plants to overcome different types of stress. It will help plants information of gametes during sexual reproduction and also protect the host plant against fungal infection by degrading its cell wall. This enzyme also acts on host cell wall-associated glucan and helps in cell expansion for its growth and also maintains cell wall fluidity under cold stress to reduce freezing damage. It has the potential to form a complex with ice crystal formed in plant cells under cold stress, so regulate the size of ice crystal for the development of cold stress tolerance. So the bacterial glucanase is considered an environmentally safe solution for many environmentally dangerous complications to increase the crop yield without chemical input in an eco-friendly manner.

7.1 Introduction

Beta-glucanase is a common pathogenesis-related protein of the plant defense system, which induces in plants in response to pathogenic interaction and is broadly distributed in higher plants. Beta-glucanase enzyme catalyzes the hydrolytic cleavage of the beta-1,3-glucans linkage which is the main structural component of the fungal cell wall and this enzyme has been produced by many microbes also, to check the growth of other organisms to reduce the competition for food and niche in its surrounding as well as to avail food from the competitor for its growth and development. This enzyme in plant serves as defense enzyme as it inhibits and blocks the growth of pathogens and the hydrolysis product of this enzyme is oligosaccharide,

Y. Jha (✉)
N.V.Patel College of pure & Applied Sciences, CVM University, Anand, Gujarat, India

N. S. Pradeep, L. K. Edison (eds.), *Microbial Beta Glucanases*, Interdisciplinary Biotechnological Advances, https://doi.org/10.1007/978-981-19-6466-4_7

which acts as an elicitor to induce a cascade of systemic resistance in plants and helps in the induction of other related PR proteins and defense-related products (Singh et al. 2014). In addition to plant protection response, beta-glucanase is involved in various plants developmental and metabolic activities like pollen germination, bud dormancy, flower growth, fruit ripening, and seed germination (Tao et al. 2013).

The plant–microbe interaction is one of the natural activities and is highly diverse, as the plant has been associated with microbes below the ground with roots, above the ground on the shooting surface as well as within the plants with various tissues. So, such interaction can be both epiphytic and endophytic, depends on the types of association and aim of the association. The interaction between microorganisms and plants may involve several mechanisms leading to either neutral, synergistic, or negative consequences which are important for both plants and microbial community and depends on types of microbes as well as competition, which is directly related with the plant growth, health, and development (Jha 2019a, b, c). During evolution, this interaction has developed specificity and particular plant species may act as host only for limited number of microbes. But still, plants regularly have deleterious microbial interaction, which leads to infection in the plants. So, plants with evolution develop multilayer mechanisms to check pathogenic interaction and respond efficiently by inducing specific defense responses. But pathogenic interaction of crop plants is a major cause for potential crop yield loss every year. Crop farming is till date only way for food production and the farming system continuously facing a great challenge due to frequent environmental changes, growing population, and industrialization. Due to intensive urbanization, there is continuously reducing agricultural land and due to continuously growing population, there is growing food demand for fulfilling the food requirement of a growing population, which is one of the major concerns to be addressed presently due to the limited natural resources. Cereals, legumes, pulses, and vegetable need to be substantially produced to fulfil the growing food requirements for the growing population of the world, but the yield of these crops are continuously challenged with poor soil quality and different types of stresses (Onaga and Wydra 2016). To meet the growing demand in food production, yield and productive potentials of major crops are attained by the exhaustive use of pesticides and fertilizers; but their application also results in diverse pollutions and contamination of ecosystems. There is need to incorporate some novel and ecofriendly tool in our agriculture practices, to increase the crop yield for addressing the future food demand for the growing population for the sustainability of our natural ecosystem and environment (Majeed et al. 2015).

The rhizosphere is a rich niche of microorganisms that interact with plants and promote different signalling processes and exchange of materials in the plant-microbe interface. Among rhizosphere microorganisms, bacteria hold a key position in the natural ecosystem as well as in agriculture because they can stimulate plant growth by fixing nitrogen, improving soil nutrients, suppressing pathogenic microbes interaction, and modifying plants' responses to biotic and abiotic stresses (Shameer and Prasad 2018). Also, such interaction has stimulatory effects on soil nutrients auction, better nutrient uptake, overall physiological activities and management of stress conditions of the plants, and induces immune responses of plants toward pathogens. Many strains of bacteria are positively linked with suppressing diverse plant pathogens by producing antagonistic metabolites and enhancing the immunity potential of crops under biotic stress (Berendsen et al. 2018).

7.2 Isolation of Microbial Having Enzyme β-1,3-Glucanase Activity

The simplest forms of life on the earth are of microorganisms but have a great impact on the other living forms. Microbes are everywhere in the biosphere, and their presence invariably affects the environment in which they are growing. The effects of microorganisms on their surrounding environment can be beneficial or harmful or neutral and depend on the activity influenced by them. The microbes have specific internal chemistry and have diverse sets of natural active enzymes capable of catalyzing diverse chemical reactions for the living world. Microbial enzymes are involved in many diverse reactions in many aspects of a living system and supporting every surrounding organism as well as agricultural production naturally (Thapa et al. 2020). Now microbial enzymes are explored for practical applications, industrial usage and development of unique drugs as well as for researches. The microbial enzymes are extensively explored for the development of eco-friendly agricultural production technology and to reduce labor input. The increasing use of microorganisms and their valuable products are contributing to eco-friendly sustainable agricultural development and is a promising tool for fulfilling the requirement of enhanced food production, without application of chemical inputs in crop farm, which have adverse affects the on the global biodiversity. The sustainability of agricultural practice requires the enhanced use of biological agents as biofertilizers and biocontrol agents, to promote organic production.

The microbes for the study has been isolated from the rhizosphere of paddy as per our established procedure and thirty-seven different isolates have been obtained, among them, two strains of *Pseudomonas pseudoalcaligenes* and *Bacillus pumilus* were selected based on its growth promotion ability and β-1, 3-glucanase activity for the further study (Jha 2017). To analyze the ability of bacteria for its β- 1, 3-glucanase production potential, all bacterial isolates are screened for their glucanolytic activity using β-glucan agar plate for culturing. Initial detection of β-glucanase was assayed by congo-red staining. A clear zone of β-glucan hydrolysis around the bacterial colony showed an indication of β-glucanase production by bacteria. The glucanolytic index is calculated using the equation: (A − B) B-1, where A is the diameter of the clear zone, and B is the diameter of bacterial colonies. The screening of bacterial β-glucanase showed that two isolates are having high glucanolytic activity, as indicated by the formation of the biggest clear zone around the bacterial colony as well as the highest glucanolytic index (Table 7.1). The clear zone formed

Table 7.1 Analysis of beta-glucanase activity in the *B. pumilus* and *P. pseudoalcaligenes* ($n = 5$)

Bacterial Isolates	Zone of Clearance (mm)	Index of Glucanolytic	β-1, 3-glucanase production (nmol min^{-1} g^{-1})
P. Pseudoalcaligenes	4.7 ± 1.12	7.23 ± 0.23	0.63 ± 0.11
B. pumilus	3.9 ± 0.22	6.14 ± 0.31	0.58 ± 0.46

around the colony showed substrate hydrolyzed into glucose. The congo-red dye will detect nonpolar hydrogen linked (β-glycosidic) on polysaccharides. The detection is due to the strong hydrophobic interactions of van der walls between dye and polysaccharide which contain nonpolar hydrogen links (Mazeau and Wyszomirski 2012).

Then ten isolates having the highest glucanolytic activity have been selected for quantitative β-1, 3-glucanase production analysis and the quantitative analysis has been done by inoculating the bacterial cultures in N-Broth and allowed to grow for 24 hrs at 30 °C on a shaker at 150 rpm. The bacterial culture suspension has been centrifuged at 10,000 g for 20 min and the supernatant has been used as an enzyme source. The reaction mixture comprised of 62.5 μl of crude enzyme extract and 62.5 μl of laminarin (4%) has been then incubated at 40 °C for 10 min. The reaction has stopped by adding 375 μl of dinitrosalicylic acid reagent and heated for 5 min in a boiling water bath. The resulting dark brown colored solution has been diluted with 4.5 ml of water and taking absorbance at 540 nm. The enzyme activity has been expressed as nmol min^{-1} mg^{-1} (Farheen et al. 2020). The results of quantitative β-1, 3-glucanase production analysis showed that the β-1, 3-glucanase production increased with time in all the isolates, but the two isolates aving the highest ability had been selected for the study (Table 7.1). The selected bacteria were identified by biochemical and molecular techniques. For molecular identification, the ribotyping has been used. The bands on the agarose gel obtained by PCR amplification of 16SrDNA with the help of 16S universal primer 8F: 50 AGAG TTTGATCCTGGCTCAG30 and 1510R: 50 GGCTACCTTG TTACGTA3 have been eluted and sequenced. The obtained nucleotide sequences of selected bacteria has been identified as *P. pseudoalcaligene* having accession Number; EU921258 and *B. pumilus* having accession Number; EU921259 after BLAST and nucleotide sequences are submitted in the NCBI data base.

7.3 Various Types of Beta-Glucanase Produced by Microbes

The enzyme glucanases are classified as O-glycosyl hydrolase family based on structural criteria and preferentially act on the glycosidic bond between two carbohydrates, or between a carbohydrate and non-carbohydrate moiety having 1,3-β-D-glycosidic linkages in (1–3)-β-D- and (1–3),(1–6)-β-D-glucans, which present in the cell walls of many pathogenic fungi. It acts on substrates having linear glucose

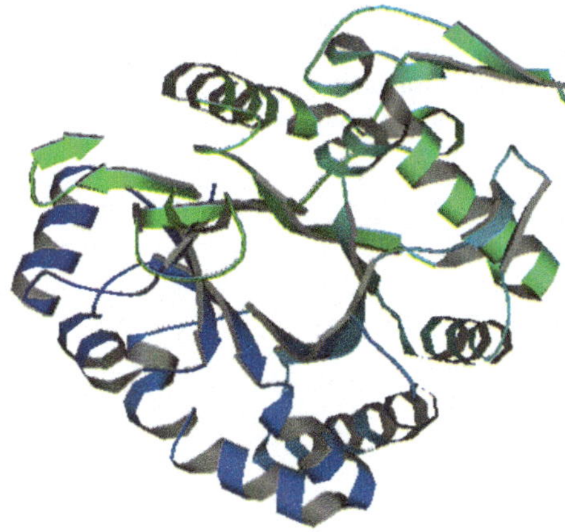

Fig. 7.1 Molecular model of endo-1,3-beta-glucanase monomer with 334 residues

arrangement linked by (1–3)-β-D- and (1–3),(1–6)-β-D linkage, where it acts by successive hydrolysis at the non-reducing end of the substrate, generating free glucose and oligosaccharides. So, enzymes β-1,3-glucanase are divided into two types endo- and exo-hydrolases, based on their action pattern (Mouyna et al. 2013). Endo-hydrolase cleaves β-glycosidic linkages at random sites laterally and releases smaller oligosaccharides, whereas exo-hydrolase catalyzes the hydrolysis of the non-reducing ends of the β-glucan chain and sequentially releases glucose residues. According to the site of action on β-glucoside linkages and hydrolytic action toward specific substrates, it is classified into numerous types:

Endo-1,3-β-Glucanase (EC 3.2.1.39)—It is also known as endo-1,3-beta-D-glucosidase, 1–3-β-glucan 3-glucanohydrolase, laminaranase, etc. and acts upon the internal 1,3-β-D-glucosidic linkages in 1,3-β-D-glucans (Fig. 7.1). It hydrolyses substrates like laminarin, curdlan and paramylon. It has functional and developmental role in plant and help in defense against phytopathogenic fungus.

Endo-1,4-β-Glucanase (EC 3.2.1.4)—It is also known as endo-1,4-β-D-glucanohydrolase, celludextrinase, endoglucanase D and carboxymethyl cellulose (Fig. 7.2). It hydrolyses specific substrates like cellulose, lichenin, cereal β-D-glucans and avicel. It helps in hydrolysis of noncellulosic or cellulose polysaccharide component of cell wall during grain germination for embryo as well as has significant role in cellulose synthesis for cell growth.

Endo-1,6-β-Glucanase (EC 3.2.1.75)—It is also known as β-1,6-glucan hydrolase, β-1,6-glucan 6-glucanohydrolase, β-1 → 6-Glucan hydrolase, endo-β1,6-glucanase, β-1,6-glucanase (Fig. 7.3). It hydrolyzes specific substrates like 1,6-oligo-beta-D-glucosides, lutean, and pustulan, one of the major cell wall components of lichens and fungus. *β*-glucanases have been involved in defense against pathogenic fungi, via disruption of their cell walls (Temple et al. 2017).

Endo-1,3 (4)-β-Glucanase (EC 3.2.1.6)—It is also known as β-1,3-glucanase, endo-β-(1–3)-D-glucanase, endo-β-1,3–1,4-glucanase, endo-β-1,3 (4)-glucanase, laminarinase as it hydrolyzes the endo (1 → 3) or (1 → 4)-linkages within the β-D-glucan (Fig. 7.4). It hydrolyzes specific substrates like lichenin, cereal

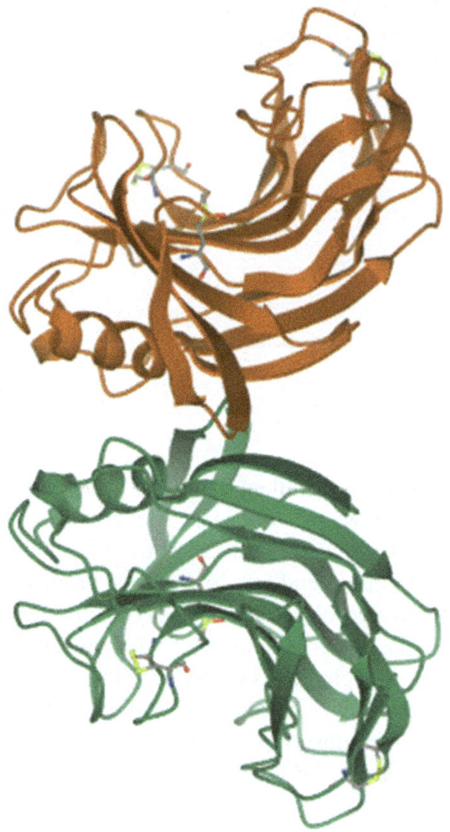

Fig. 7.2 Molecular model of endo-1,4-beta-glucanase monomer of Streptomyces lividans with 222 residues

D-glucans, and laminarin. It helps in releasing cellotetraose from lichenin and can lyse fungal cells to check fungal infection.

Endo-β-1,3–1,4-Glucanase (EC 3.2.1.73)—It is also known as 1,3; 1,4-β-glucan, endo-β-1,3–1,4 glucanase, 1,3,1,4-β-glucan endohydrolase, 1,3–1,4-β-D-glucan 4-glucanohydrolase, licheninase, as it hydrolyzes specific substrates like cereal β-glucans and lichenin only, but not on β-glucans consisting 1,3- or 1,4-bonds individually (Fig. 7.5). It is the main components of nutritional component of the endosperm and cell wall and is extensively present in algae, plants, fungi, and bacteria.

Exo-1,3-β-Glucanase (EC 3.2.1.58)—It is also known as β-1,3-glucan exo-hydrolase, exo-1,3-β-glucosidase, exo-β-1,3- D-glucanase, 1,3-β-glucan glucohydrolase, 1,3-β-glucosidase, as it hydrolyzes specific non-reducing tails of (1 → 3)-β-D-glucans to librate α-glucose (Fig. 7.6). It has important agricultural and industrial applications, including biomass conversion for biofuel or food production and aroma enhancement (Patrick et al. 2010).

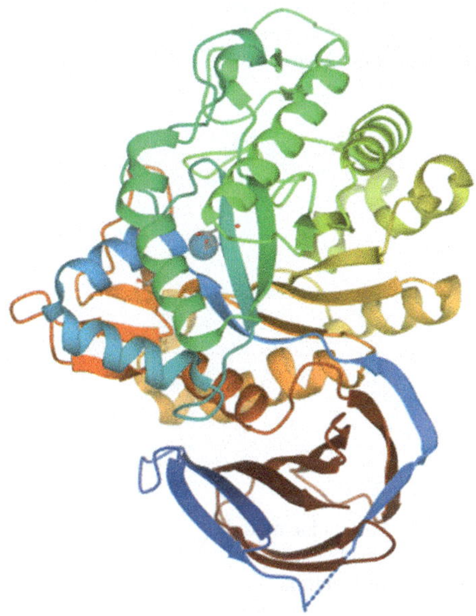

Fig. 7.3 Molecular model of endo-1,6-beta-glucanase monomer of Bacteroides thetaiotaomicron with 1356 residues

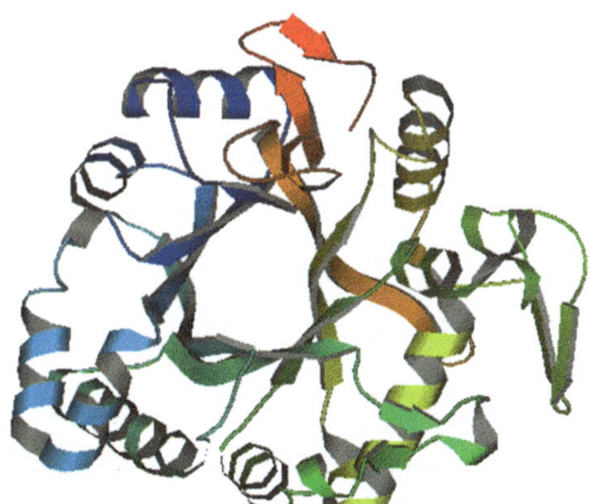

Fig. 7.4 Molecular model of endo-1,3 (4)-β-glucanase monomer with 312 residues

Exo-1,4-β-Glucanase (Ec 3.2.1.91)—It is also known as 1,4-β-glucan cellobiosidase, exo-cellobiohydrolase, exoglucanase; exo-β-1,4-glucan cellobiohydrolase, exo-1,4-β-D-glucanase,which hydrolyses specific (1→4)-β-D-glucosidic bonds from the ends of cellulose chains resulting with the release of oligosaccharides, such as cellobiose units from the polysaccharide (Fig. 7.7).

So, this enzyme plays a very versatile role and it plays a critical role in plant defense by delaying the growth of pathogenic fungi, to protect the plant from damage caused by a pathogenic microorganism, as pathogenic microorganism cell walls contain β-glucans. Apart from this, β-1,3-glucanase is also involved in diverse physiological and developmental processes such as endosperm formation, somatic embryogenesis, microsporogenesis, pollen development, seed germination, flower

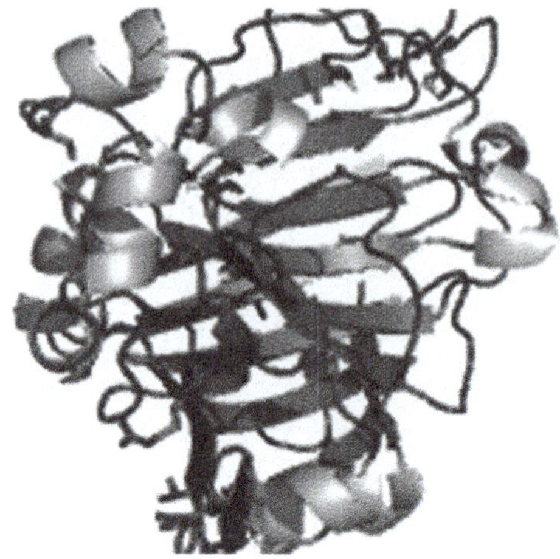

Fig. 7.5 Molecular model of endo-β-1,3–1,4-beta-glucanase monomer

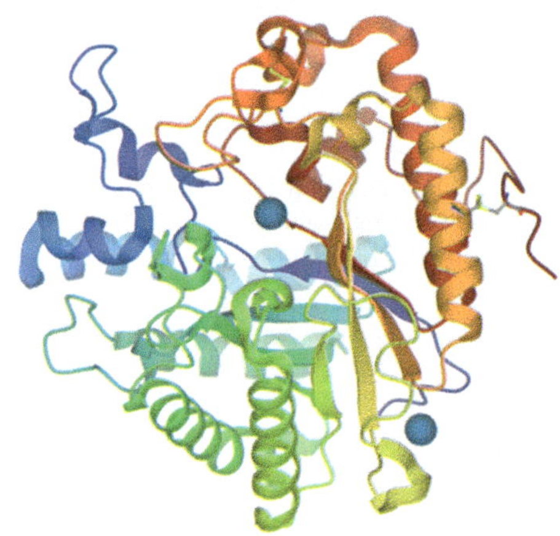

Fig. 7.6 Molecular model of exo-1,3-β-glucanase monomer of Candida albicans with 394 residues

formation, and the response to wounding and abiotic stress. The subcellular localization of hydrolytic enzymes in plants can provide important information about the role of hydrolytic enzymes during defense against pathogens (Yarullina et al. 2016).

7.4 Effect of Microbial Beta-Glucanase on Embryogenesis, Seed Germination, and Dormancy

During evolution, plants develop many innovative mechanisms for their better survival in changing environmental conditions, and among all such innovations ability to form seeds by the vascular plant is one of the critical innovations. Seed acts as a reproductive unit in the vascular plant and serves as a connecting link for successive generations. It protects the plant embryo from the external environment

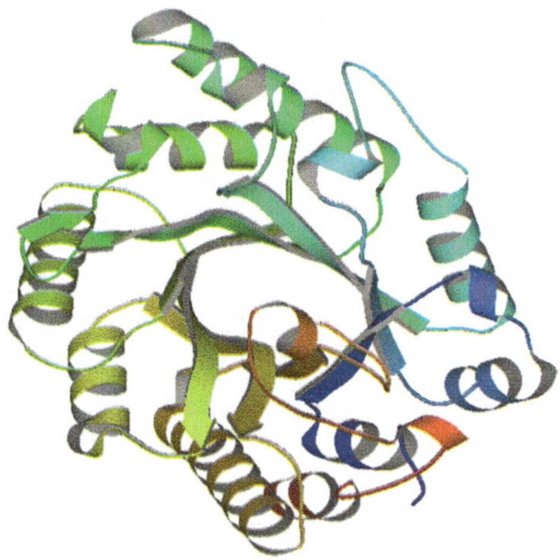

Fig. 7.7 Molecular model of exo-1,4-β-glucanase monomer with 312 residues

and helps in its dispersal from one location to other to reduce inter- and intraspecies competition. The seeds are biochemically and physiologically completely different from the plant and extremely important for the existence of a wild plant species in the changing environment and for storing plant embryos for the next germination season (Zhang et al. 2007). Structurally seed shows vast diversity, but basically, it is the same and contains only three components: seed coat, endosperm, and embryo. In which embryo and endosperm are the product of double fertilization and the outer covering seed coat developed from ovular tissues.

The embryogenesis in plants initiates with double fertilization, which results in the formation of endosperm and embryo, which remain protected in the form of seed. The development of a new plant takes place by the process of a complex mechanism of embryogenesis. The plant cell wall is composed of different types of carbohydrates and proteins and forms a complex network, which remains around the embryo for its protection. The callose component of the cell wall needs to degrade gradually for the growth of the embryo and finally deposited callose completely hydrolyzed by β-1,3-glucanases, to be used by the growing embryo (Zavaliev et al. 2011). So for somatic embryogenesis also the genes for corresponding enzymes have been used and expressed in coordination as the choice method. By the β-1,3-glucanases activity on the cell wall not only degrade glycoprotein/carbohydrate to make pore for the emergence of the embryo from seed, but it produces oligosaccharides of carbohydrate or glycoprotein as a byproduct of cell wall callose hydrolysis, which serves as a source of nutrition of the developing structures, in this way enzyme activity regulates the process of the embryogenesis as per the surrounding environment.

The seed coat due to the presence of a thick layer of β-1,3-glucan is tough to protect the embryo from mechanical injury and is semi-permeable for the exchange of oxygen and water. The seed coat acts as a physical barrier for radicle emergence and is responsible for seed dormancy. The expression of β-1,3-glucanase in the seed has a significant role in the regulation of seed germination, dormancy, and protection against the seed pathogen. The seed germination starts with imbibitions of water,

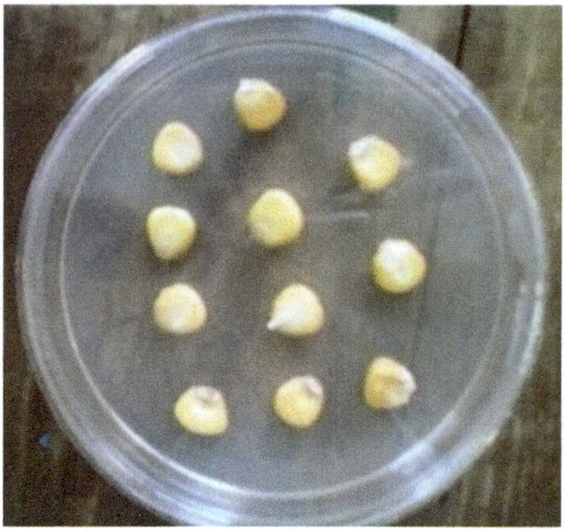

Fig. 7.8 Plate showing germination of maize seed on β-glucan agar medium

results in the expansion of the embryo and finally it requires rupturing of the seed coat for the emergence of the radical. During seed germination endosperm weakening and dissolution is necessary for successful completion of the emergence of the radical requiring the activity of the beta-glucanase enzyme (Steinbrecher and Leubner-Metzger 2017). In the present study, the effect of microbial glucanase has been analyzed on maize seed germination. Seeds of maize variety Pioneer 30 V92 have been washed thoroughly with distilled water followed by surface sterilization with 0.1% $HgCl_2$ solution for 4 min and 70% ethanol for 10 min. The seeds are washed thoroughly with sterile distilled water and kept on a shaker for 5–6 hrs in sterile distilled water on a rotary shaker. Later the seeds are transferred to Petri dishes containing β-glucan agar medium containing individual bacterial isolate *P. pseudoalcaligene* and *B. pumilus* and incubated in dark at 30 °C (Fig 7.8). The germination parameters like imbibition period, total germination period, germination index, germination value, mean germination time, and speed of emergence have been analyzed. The result of the study showed that isolated bacteria *P. pseudoalcaligene* and *B. pumilus* both have a significant effect on the seed germination parameter of maize. The imbibition and total germination period are remarkably shortened, while the percentage of seed germination and seed germination index remarkably increased in presence of *P. pseudoalcaligene* and *B. pumilus* glucanase compared to the control. The mean germination time significantly reduced, but the rate of seed emergence gradually increased as shown in Table 7.2. So the regulation of seed germination and dormancy is directly related to the production of the β-glucanase enzyme. The induction of β-1,3-glucanase takes place within the micropylar endosperm to degrade cell wall material just before it ruptures for proper germination in response to plant hormone-like abscisic acid, ethylene, gibberellins, and sunlight as reported by Leubner-Metzger (2003). As the enzyme β-1,3-glucanase digesting β-1,3-glucan callose layer of seed, which is located between the plasma membrane and the cell wall of seed are directly responsible for seed germination.

Table 7.2 Analysis of effect of beta-glucanase activity of the *B. pumilus* and *P. pseudoalcaligenes* on the maize seed germination parameters ($n = 5$)

Bacterial isolates	Imbibition period	Total germination period	Germination %	Germination index	Germination value	Mean germination time	Speed of emergence
Control	67 ± 0.54	94 ± 0.31	86 ± 1.21	5.62 ± 0.4	5.62 ± 1.22		
P. Pseudoalcaligenes	56 ± 0.21	82 ± 0.22	94 ± 0.32	7.23 ± 0.2	6.77 ± 0.11	79.32 ± 0.21	41.2 ± 0.13
B. pumilus	48 ± 0.27	74 ± 0.41	89 ± 0.12	6.14 ± 0.1	5.89 ± 0.26	74.87 ± 0.42	38.7 ± 0.23

7.5 Effect of Microbial Beta-Glucanase in Cell Division

The plant cell remains surrounded by a complex cell wall structure and is responsible for a wide array of functions in plant life. It helps plant cells in resisting internal hydrostatic pressures to maintain the structural integrity of plant cells as well as serve as a physical barrier against pathological and environmental stress. It helps in cell division by providing flexibility to the plant cell and enabling cell differentiation by providing a suitable biochemical scaffold (Tucker and Koltunow 2014). The plant cell wall is critical for the normal growth and development of plants and it carries a broad range of receptors, channels, and pores, which regulate the molecular communication of proteins, enzymes, sugars, signals, and hormones as well as protect the protoplast and provide rigidity to the cellular structure of the plant. It has many more functions like cell shape determination, cell adhesion, protoplast protection, maintaining turgor pressure and cell water balance, mediating cell–cell interactions and regulation of the macromolecules diffusion. The plant cell wall structure is extremely variable among different plant species and also among different tissue types in the same species. For the proper growth of the plant, the synthesis of new cells between two cells initialized with the development of a new wall during cell division, followed by cell expansion. The cell expansion directly depends on cell water uptake, separation of cellulose microfibrils, and addition of new cell wall polysaccharides.

The typical cell wall constituent includes proteins, phenolic compounds, cellulose, non-cellulosic, and pectic polysaccharides and among which polysaccharides are the major constituent (Kumar et al. 2016). In which protein serves as cellular receptors, phenolic compounds protect from a pathogen, non-cellulosic polysaccharides including β-1,3:1,4-glucan, callose, xyloglucan, and mannan provide mechanical support, while pectin provides strength, adhesion, porosity, acts as intercellular signalling and helps in expansion by forming a complex network with proline-rich glycoprotein extension. The enzyme endo-1,4-β-D-glucanases plays a remarkable role in cell expansion by cleaving β-1,4-glycosidic bonds of β-1,3:1,4-glucan, callose, and xyloglucan. Microbial endo-beta-glucanases contain a cellulose-binding domain for the degradation of cellulose to its monomer to be used as a carbon source for its establishment in host cell and growth. In the present study, the effect of microbial glucanase has been analyzed on the maize cell wall expansion and cell division in the form of early plant growth. Certified seeds of maize variety Pioneer 30 V92 are obtained from the Gujarat state seed cooperation and are inoculated with isolated bacteria by transferring the germinating seed to culture tubes containing 400 μl Hoagland's nutrient medium, 400 μl micronutrients, 400 μl β-glucan, and 1% agar in 40 ml distilled water. Before the transfer, the bacterial inoculum of the isolated bacteria was added to the medium at a concentration of 6×10^8 CFU ml^{-1}. To obtain a mixture of both bacterial cultures, an equal volume of both the cultures were mixed in the medium to give a concentration of 6×10^8 CFU $ml^{-1}$and all the tubes are incubated at 27 °C in a 12 hrs light-dark cycle in a growth chamber. The result of the study showed that with the hydrolysis of β-glucan (reduced turbidity) direct increase in the growth of plant seedlings has been observed (Table 7.3).

7.6 Effect of Microbial Beta-Glucanase in Defense Against Pathogen

Plants induce diverse methods for their protection/to activating specific defense responses, to act properly upon pathogenic microbial interaction. Plant activates diverse mechanisms that start with the activation of primary barriers like cell wall reinforcement with antimicrobial peptides, enzymes, non-proteinaceous secondary metabolites, lignin, suberin, and callose. After primary barriers are invaded by the pathogen, the plant activates a cascade of defense responses like enhanced deposition of lignin, suberin, and callose for rigidification of cell wall, activation of hypersensitive reaction, induction of pathogenesis-related proteins, production of phenolic compounds, phytoalexins, and reactive oxygen species collectively known as "defense responses" in a plant. Besides having a vast morphological difference in the plant, all plant activates the production of structurally and functionally basic protective proteins for its protection against phytopathogens known as pathogenesis-related (PR) proteins. PR proteins are sturdily induced upon infection by pathogens/wounding and rapidly accumulate at the localized site of infection, results in the induction of systemic acquired resistance (Jha 2019a, b, c). The microbial β-1,3-glucanase hydrolyzes β-1,3-linked glucans layer of the plant cell wall during infection for its invasion and to acquire food, results in the production of oligosaccharides, which act as an elicitor. The microbial β-1,3-glucanase hydrolyzes β-1,3-linked glucans, the major structural component of fungal cell walls and is responsible for the induction of plant defense mechanism against fungal infection. The first step in induction of plant defense mechanism is the interaction of the elicitor molecules with putative receptor of the plant defense system. And it is significantly necessary to characterize the structural elements configuration to be recognized by the receptor molecules, which is the molecular basis for the perception of elicitor by the receptors for the transduction of elicitor signals in plants (Jha 2019a, b, c). In the present study, the effect of isolated bacterial glucanase in the induction of plant defense by inducing β-1, 3-glucanase activity has been analyzed in the maize by using certified seeds of maize variety Pioneer 30 V92. The maize plant has been allowed for root association with selected isolated *P. pseudoalcaligenes* and *B. pumilus* bacteria as above said method. After two weeks the plant leaves (5 g) are homogenized in 5 ml of 0.1 M borate buffer, pH 7.0 containing 0.1 g polyvinylpolypyrrolidone (PVP) and the supernatant has been used for enzyme assays. β-1, 3-glucanase activity in the inoculated plant has been assayed by the laminarin-dinitrosalicylic acid method. The enzyme activity was expressed as 1 nmol of reducing substances min^{-1} mg^{-1} of fresh weight. The result of the study showed that β-1, 3-glucanase activity increases with time in the plants inoculated with

Table 7.3 Analysis of effect of beta-glucanase activity of the *B. pumilus* and *P. pseudoalcaligenes* on the maize plant growth parameters ($n = 5$)

Bacterial isolates	Shoot length (cm)	Root length (cm)	Total plant height (cm)	Shoot weight (g)		Total Biomass (g)	
				Fresh	Dry	Fresh	Dry
Control	15.23 ± 0.11	12.87 ± 0.21	28.10 ± 0.27	1.83 ± 0.1	0.171 ± 0.1	2.73 ± 0.5	0.23 ± 0.1
P. Pseudoalcaligenes	24.54 ± 0.25	19.21 ± 0.16	43.75 ± 0.22	2.67 ± 0.3	0.312 ± 0.3	3.64 ± 0.1	0.32 ± 0.2
B. pumilus	22.12 ± 0.31	18.76 ± 0.24	40.98 ± 0.13	2.43 ± 0.4	0.298 ± 0.1	3.32 ± 0.2	0.29 ± 0.1

Table 7.4 Analysis of effect of beta-glucanase activity of the *B. pumilus* and *P. pseudoalcaligenes* on the β-1, 3-glucanases induction, sugar accumulation and cell membrane stability in maize leaves ($n = 5$)

Bacterial isolates	β-1, 3-glucanases (nmol of glucose min^{-1} g^{-1})	Sugar concentration (%wt/wt)	Cell membrane stability
Normal			
Control	0.32 ± 0.22	18.1 ± 0.16	123.1 ± 0.41
P. Pseudoalcaligenes	0.52 ± 0.14	27.4 ± 0.22	147.3 ± 0.01
B. pumilus	0.48 ± 0.12	26.3 ± 0.31	149.2 ± 0.31
Stressed			
Control	0.42 ± 0.11	21.6 ± 0.31	118.2 ± 0.22
P. Pseudoalcaligenes	0.58 ± 0.13	29.31 ± 0.11	135.2 ± 0.41
B. pumilus	0.62 ± 0.23	30.11 ± 0.26	132.8 ± 0.02

bacterial isolate either alone or in combination. The highest β-1, 3-glucanase activity is shown by the plant inoculated with *P. pseudoalcaligenes,* followed by plants inoculated *B. pumilus* compared to non-inoculated control plants in absence of pathogen (Table 7.4). The effect of bacterial glucanase has further been analyzed in inoculated maize in presence of fungal pathogen *Aspergillus niger,* for that 4 weeks old inoculated plants are infected with *Aspergillus niger* spore suspension with a spore load of 10^{-4} conidia ml^{-1}, which caused more than 75% infection under greenhouse condition. Then infected plants are transplanted in a pot, having five transplants per pot holding 10 kg soil and allowed to grow without additional nutrients for 1 week at 20 to 25 °C in the greenhouse. The results of the study showed enhanced β-1, 3-glucanase activity in plants inoculated with selected bacteria in presence of the fungal pathogen, compared to non-inoculated control plants, which better help plants to develop resistance against phytopathogen. The induction of beta-1,3-glucanase genes has been regulated by some environmental stress also like salinity, plant hormone and pathogen infection (Su et al. 2016). So, the production of cell wall lytic enzymes and induction of host resistance are considered to be important actions of microbial biocontrol agents against pathogenic fungi.

7.7 Effect of Microbial Beta-Glucanase in Defense in Cold Stress Tolerance in Plant

Plants are forced to get exposed to different environmental stress during their lifespan. The two main types of stress are biotic stress which include infection by pathogenic microbes viruses, nematodes, bacteria, fungi, etc. and abiotic stress which include different types of environmental unfavorable conditions as nutrient deficiency, salinity, drought, cold, heat, flood, etc. All such stresses harm crop survival and productivity. Plant evolved many physiological, biochemical, and morphological mechanisms to develop tolerance against such stress factors for their better survival. The earliest noticeable effect of stress is retarded plant growth due to retarded cell extension. The plant cell shape, size, growth/expansion are mechanically controlled by a plant cell wall. It is a composite arrangement of non-cellulosic neutral polysaccharides and cellulose microfibrils cross-linked with structural proteins and present in a physiologically active pectin matrix with lignin, which acts as a protective barrier of the cell against pathogen invasion as well as changing environmental stimulus (Jha 2017). The cell wall extensibility is regulated by cell wall-related proteins, for the desired cell expansion and enlargement. Such cell wall-related proteins includes expansins, endo-1,4-β-D-glucanase, and xyloglucan endo-β-transglucosylase are responsible for maintaining cell wall plasticity.

A major abiotic stress responsible for limited productivity and growth of agricultural crops is low temperature, parallelly the plants evolved many competent mechanisms at biochemical and physiological level to adapt in the low temperature, which induce specific enzyme activity related with cell wall modification, accumulation of osmoprotectants, sugars and induction of production unsaturated fatty acid to main tain cell fluidity. The metabolic changes concerning protein and sugar content in plants take place for cold acclimation and the accumulation of small protein and sugar molecules in plant cells helps in the development of cold tolerance. It would help in minimizing the osmotic effect and ice formation under freezing-induced dehydration conditions. The enzyme *β*-1,3-glucanases act on high-*molecular-weight* sugar into smaller sugar for osmoprotective effect as well as act on cell wall glucans to maintain cell fluidity to reduce the brittleness of the cell. Another adverse effect of cold temperatures is freezing injury of the plant cell, which increases the plant's cell susceptibility for pathogen attack. Whereas most of the pathogens are less active and virulent, so unable to attack plants at low temperatures. But psychrophilic pathogens can invade plants under this condition and are responsible for remarkable crop yield loss (Jha and Yadav, 2021). The *β*-1,3-glucanase enzyme also acts on the fungal cell wall to check its infection, as its cell wall contains *β*-1,3-glucan chains. With all these activity glucanases enzyme also can adsorb to ice crystal surfaces and regulate the growth of ice crystal, affecting its hydrolytic activity. Glucanase enzymes have the structural advantage of having separate hydrolytic catalytic and ice-binding domains. In the present study effect of microbial β-1,3- glucanase has been analyzed on cell membrane stability and in the accumulation of sugar under cold stress. Cell membrane stability in leaves was determined in new leaves per treatment and total soluble sugar is measured spectrophotometrically using the method of Ibrahim and Jaafar (2012). Water affects the plant's growth and development and decreasing external water potential produces a net accumulation of solutes in cells, which lowers the cell osmotic potential necessary for maintaining the turgor pressure. The result of the study showed that plants inoculated with isolated bacteria under cold stress have higher cell membrane stability as well as sugar concentration compared to non-inoculated control. So the finding suggests that the microbial-induced *β*-1,3-glucanases in the maize plant have a dual defense function simultaneously against both biotic and osmotic stress under cold conditions.

7.8 Conclusion

The agriculture sector is adversely affected by the different environmental stress factors, which results in huge economic losses. Agricultural crop production also suffered due to different phytopathogenic interactions, but plants also have beneficial interactions with microbes, which help plants in their better survival. Some plant-associated bacteria have the potential for the production of beta-glucanase enzyme and this microbial glucanase coupled with host glucanase and increase the efficiency of enzyme-related activity like seed germination, breaking seed dormancy, gamete formation, and check pathogen infection. It also helps host plants under abiotic stress by degrading large sugar into smaller ones to maintain cell osmotic levels as well as act on plant cell wall sugar to reduce cold freezing damage. So microbial enzymes decrease environmental degradation for sustainable agriculture practice and act as an economically feasible tool for increasing productivity and food safety in place of chemicals.

References

Berendsen RL, Vismans G, Yu K, Song Y, de Jonge R, Burgman WP, Burmølle M, Herschend J, Bakker PAHM, Pieterse CMJ (2018) Disease-induced assemblage of a plant-beneficial bacterial consortium. ISME J12:1496–1507

Farheen N, Najm-ul-Seher MY, Khan MJ, Nadeem SM, Ahmad M (eds) (2020) Soil microbes and plant health. Plant disease management strategies for sustainable agriculture through traditional and modern approaches, pp 111–135

Ibrahim MH, Jaafar HZ (2012) Impact of elevated carbon dioxide on primary, secondary metabolites and antioxidant responses of Eleais guineensis Jacq. (oil palm) seedlings. Molecules 17 (5):5195–5211

Jha Y (2017) Cell water content and lignification in maize regulated by rhizobacteria under salinity. Brazilian Journal of Biological Sciences 4(7):9–18

Jha Y (2019a) Endophytic bacteria mediated anti-autophagy and induced catalase, β-1, 3-glucanases gene in paddy after infection with pathogen *Pyricularia grisea*. Indian Phytopathology 72(1):99–106

Jha Y (2019b) Higher induction of defense enzymes and cell wall reinforcement in maize by root associated bacteria for better protection against *aspergillus Niger*. J Plant Prot Res 59(3): 341–349

Jha Y (2019c) Endophytic bacteria as a modern tool for sustainable crop management under stress. In: Giri B, Prasad R, Wu QS, Varma A (eds) Biofertilizers for sustainable agriculture and environment. Soil biology, vol 55, pp Springer–SpriCham

Jha Y, Yadav AN (2021) Piriformospora indica: biodiversity, ecological significances, and biotechnological applications for agriculture and allied sectors. In: Abdel-Azeem AM, Yadav AN, Yadav N, Usmani Z (eds) Industrially important fungi for sustainable development. Fungal biology. Springer, Cham

Kumar M, Campbell L, Turner S (2016) Secondary cell walls: biosynthesis and manipulation. J Exp Bot 67(2):515–531

Leubner-Metzger G (2003) Functions and regulation of β-1,3-glucanase during seed germination, dormancy release and after-ripening. Seed Sci Res 13:17–34

Majeed A, Abbasi MK, Hameed S, Imran A, Rahim N (2015) Isolation and characterization of plant growth-promoting rhizobacteria Author's personal copyPlant cell from wheat rhizosphere and their effect on plant growth promotion. Front Microbiol 6:198

Mazeau K, Wyszomirski M (2012) Modelling of Congo red adsorption on the hydrophobic 510 surface of cellulose using molecular dynamics. Cellulose 19:1495–1506

Mouyna I, Hartl L, Latgé JP (2013) β-1,3-glucan modifying enzymes in *aspergillus fumigatus*. Front Microbiol 4:81

Onaga G, Wydra K (2016) Advances in plant tolerance to abiotic stresses. In: Abdurakhmonov IY (ed) Plant genomics. InTech, Rijeka. https://doi.org/10.5772/64350

Patrick WM, Nakatani Y, Cutfield SM, Sharpe ML, Ramsay RJ, Cutfield JF (2010) Carbohydrate binding sites in Candida albicans exo-beta-1,3-glucanase and the role of the Phe-Phe 'clamp' at the active site entrance. FEBS J 277:4549–4561

Shameer S, Prasad T (2018) Plant growth promoting rhizobacteria for sustainable agricultural practices with special reference to biotic and abiotic stresses. Plant Growth Regul 84:603–615

Singh D, Ambroise A, Haicour R, Sihachakr D, Rajam MV (2014) Increased resistance to fungal wilts in transgenic eggplant expressing alfalfa glucanase gene. Physiol Mol Biol Plants 20:143–150

Su Y, Wang Z, Liu F, Li Z, Peng Q, Guo J, Xu L, Que Y (2016) Isolation and characterization of ScGluD2, a new sugarcane beta-1,3-Glucanase D family gene induced by Sporisorium scitamineum, ABA, H2O2, NaCl, and CdCl2 stresses. Front Plant Sci 7:1348. https://doi.org/10.3389/fpls.2016.01348

Tao YX, Xie BG, Yang ZY, Chen ZH, Chen BZ, Deng YJ (2013) Identification and expression analysis of a new glycoside hydrolase family 55 exo-β-1, 3-glucanase-encoding gene in *Volvariella volvacea* suggests a role in fruiting body development. Gene 527:154–160

Temple MJ, Cuskin F, Basle A, Hickey N, Speciale G, Williams SJ, Gilbert HJ, Lowe EC (2017) A Bacteroidetes locus dedicated to fungal 1,6-beta-glucan degradation: unique substrate conformation drives specificity of the key endo-1,6-beta-glucanase. J Biol Chem 292:10639–10650

Thapa S, Mishra J, Arora N (2020) Microbial cellulolytic enzymes: diversity and biotechnology with reference to lignocellulosic biomass degradation. Rev Environ Sci Biotechnol 19:621–648

Steinbrecher T, Leubner-Metzger G (2017) The biomechanics of seed germination. J Exp Bot 68: 765–783

Tucker MR, Koltunow AMG (2014) Traffic monitors at the cell periphery: the role of cell walls during early female reproductive cell differentiation in plants. Curr Opin Plant Biol 17:137–145

Yarullina LG, Akhatova AR, Kasimova RI (2016) Hydrolytic enzymes and their proteinaceous inhibitors in regulation of plant–pathogen interactions. Russ J Plant Physiol 63:193–203

Zavaliev R, Ueki S, Epel BL (2011) Biology of callose (β-1,3-glucan) turnover at plasmodesmata. Protoplasma 248:117–130

Zhang S, Hu J, Zhang Y, Xie XJ, Knapp A (2007) Seed priming with brassinolide improves lucerne (*Medicago sativa* L.) seed germination and seedling growth in relation to physiological changes under salinity stress. Aust J Agric Res 58:811–815

Chapter 8
Exploitation of Actinobacteria for Beta-Glucanolytic Enzymes—Screening and Characterization

Lekshmi K. Edison, S. Anu, and N. S. Pradeep

Abstract The ever increasing demand for industrially important enzymes has encouraged a search for economically important potential enzymes from the microbial world which offers ease of growth, less nutritional requirements and inexpensive downstream processing. Amongst the microorganisms, actinobacteria have an extensive range of biological activities. The actinobacterial enzymes have special interest due to their high yield, susceptibility to genetic manipulation, eco-friendliness and cost-efficiency. Hence, it offers an alternating green solution to chemical technology. The use of actinobacterial enzymes in wine and beer production has been experienced for several years. Beta-glucanase is a valuable enzymatic tool for beta-glucan depolymerization in breweries and animal feed industries. Actinobacteria is a promising candidate for the production of glucanolytic enzymes. This chapter explains the practice of novel actinobacterial stains for beta-glucanase production from under-exploited or unexplored environments. It is tremendously crucial, which unlocks exhilarating avenues in biotechnology research.

Keywords *Streptomyces* · Isolation and screening · Polyphasic characterization

8.1 Introduction

One of the primary taxonomic units, the phylum Actinobacteria, is now documented as major lineage within the Bacterial domain (Ludwig et al. 2012). The majority of actinobacteria are free-living and broadly dispersed in aquatic and terrestrial environments (Edison and Pradeep 2020a; Macagnan et al. 2006). Conventionally, actinobacteria is considered a midway form of bacteria and fungi. Hence, it is also

L. K. Edison
Rajiv Gandhi Centre for Biotechnology (RGCB), Thiruvananthapuram, Kerala, India

S. Anu
CEPCI Laboratory and Research Institute, Kollam, Kerala, India

N. S. Pradeep (✉)
KSCSTE-Malabar Botanical Garden and Institute for Plant Sciences, Kozhikode, Kerala, India

N. S. Pradeep, L. K. Edison (eds.), *Microbial Beta Glucanases*, Interdisciplinary Biotechnological Advances, https://doi.org/10.1007/978-981-19-6466-4_8

known as ray fungi. The name actinomycetes originated from different Greek words *aktis* or *aktin* (means ray) and *mukes* (means fungi). The actinomycetes hyphae raise by the combination of tip extension and branching. Superficially, they resembling as filamentous fungi because it produces mycelium and replicates through sporulation. However, similar to bacteria, the actinomycetes cells are thin with prokaryotic nucleoid and the cell wall contain peptidoglycan. Furthermore, cells are susceptible to antibiotics (Barka et al. 2016). The more satisfactory exclusive features like bacteria, delineate the phylum Actinobacteria into 'Kingdom bacteria'. Actinomycetes are Gram-positive, spore-producing, aerobic bacteria containing GC-rich genome and also have non-septate characteristic substrate and aerial mycelium. They are the most ubiquitous bacterial groups largely disseminated in nature such as soils, air, fresh water and also in saltwater (Srinivasan et al. 1991). The majority of actinomycetes are soil inhabitants (Kuster 1968), even though they are roughly distributed in various risky habitats including Antarctican cryophilic soils (Moncheva et al. 2002) deep-sea vent sediments (Colquhoun et al. 1998), deepest of Mariana Trench (Pathom-Aree et al. 2006), and also in deserts (Diraviyam et al. 2011). Most of them are found in the upper layers of alkaline soil comprising rich organic matter, and slowly decrease when increasing the depth of soil (Takahashi and Omura 2003).

The Actinobacterial phylum is one of the largest taxonomic units and contains 5 subclasses, 6 orders and 14 suborders included in the fourth and fifth volumes of *Bergey's Manual of Determinative Bacteriology* (Ludwig et al. 2012). All actinomycetes are encompassed in the order Actinomycetales and separated into four families—Actinomycetaceae, Streptomycetaceae, Mycobacteriaceae and Actinoplanaceae (Williams et al. 1989). The two commonly well-defined actinobacterial genera are Streptomyces and Micromonospora. The genus Streptomyces is immensely recognized as the principal reservoir of natural bioactive compounds such as enzymes, antibiotics, antitumor agents, and immunosuppressive agents (Terkina et al. 2006).

8.2 Actinobacterial Genus Streptomyces

Streptomyces is one of the largest genera that comes under the family Streptomycetaceae which was first announced in 1943 by Waksman and Henrici (Williams et al. 1983). Based on the number and species diversity, Streptomyces implies the principle taxonomic unit in actinomycetes (Bhattacharyya 1998). They are Gram-positive, aerobic, non-Acid-Fast bacteria containing around 70% GC content in the genome (Dehnad et al. 2010). At present, more than 500 species of the Streptomyces genus have been well-defined (Mohanraj and Sekar 2013) and also expose wide-ranging phylogenetic spread (Aderem 2005). Streptomyces is the solitary morphologically assorted Actinobacteria and found abundant in soil habitats, performing a foremost part in carbon recycling by producing a variety of hydrolytic exoenzymes. Due to the manufacture of the enormous variability of secondary metabolites, Streptomyces is considered the topmost skilled chemist in Nature with unlimited interest in medicine and industry (Hopwood 2007).

Streptomyces usually have small spores around or less than 1 μm diameter in size (Willemse et al. 2011), formed as straight, helical, or wavy spore chains (Chater 1993). Normally they have slow-growing colonies and produce an earthy odour due to the production of geosmin, a volatile metabolite. Initially, the colonies develop as smooth surfaces and later they form aerial mycelium with granular, powdery, floccose or velvety texture. They are non-motile and secrete a variety of pigments which determines the colour of vegetative and aerial mycelia. Streptomyces species are catalase-positive, nitrates reducing and cell wall comprises a large amount of L-diaminopimelic acid (L-DAP) (Hasani et al. 2014). They are chemoorganotrophic microorganisms that need organic carbon and nitrogen sources along with mineral salts as growth factors (Küster and Williams 1964). Mostly they are grown at temperatures between 10 and 37 °C and pH 6.5–8.0, but some thermophiles grow at 45–55 °C, for example, *Streptomyces thermovulgaris* and *S. thermonitrificans* (Srivibool et al. 2004). They are highly resistant to drought conditions, where it forms arthrospore. It requires less moisture to grow than other microbes (Subbarao 1999). In the biotechnological approach, Streptomyces is extremely important because they are produced two-thirds of all industrially important antibiotics (Champness 2000).

8.2.1 Streptomyces: Developmental Cycle

Streptomyces spp. characterized by a multifarious development cycle, comprises differentiation and programmed cell death (PCD). The life cycle begins with the germination of spores and the emergence of germ tubes by tip elongation (Fig. 8.1). Then the germ tube extended and developed a network of branched multinucleated hyphae, called substrate mycelium. After several days substrate mycelium starts to differentiate and extended to air forms multinucleated aerial hyphae. The *bld* genes are required for aerial mycelium development. The expression of *whi* gene causes the growth and coiling of aerial hyphae. Finally, the growth stops and it undertakes massive septation which gives rise to uninucleoid compartments. Then each compartment develops into spores and creates spore chains (Li et al. 2016; Manteca and Sanchez 2010). At the early stages of development, the substrate mycelium shows a dynamic primary metabolism and it undergoes the first round of PCD. The second round of PCD is noted in the aerial mycelium during spore formation. The cellular matters released during PCD generates a nutrient pool that supports further growth of the organism (Filippova and Vinogradova 2017; Manteca et al. 2008). Most Streptomyces strains in liquid cultures do not sporulate. Regardless of this, the industrial production of secondary metabolites is accomplished in liquid cultures. Streptomyces in submerged cultures mainly produces four morphological forms—pellets, clumps, branched hyphae, and non-branched hyphae (Pamboukian et al. 2002).

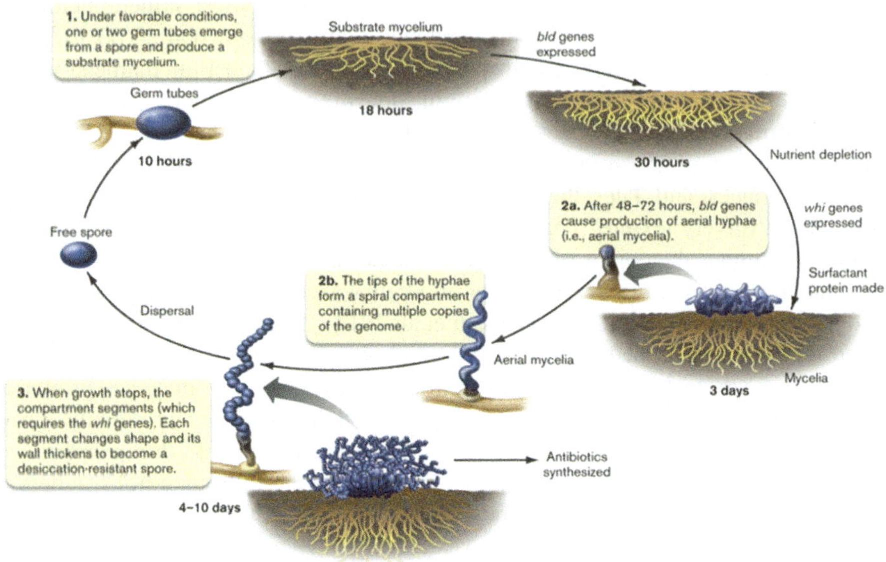

Fig. 8.1 Time course development of Streptomyces species

8.2.2 Streptomyces: A Source of Industrial Enzymes

The actinobacterial members exclusively Streptomyces are the leading and propitious manufacturers of uncountable secondary metabolites, antibiotics, and industrial enzymes. They are biologically transforming and converting many organic compounds and waste materials from agricultural and urban areas into economically value-added products. Streptomyces is the most the promising reservoir of a miscellaneous collection of economically and industrially important enzymes. Streptomyces spp. are energetically participated in the putrefaction of several organic compounds and the produced enzymes are extensively exploited in food, detergent, leather, textile, medical, and pharmaceutical industries (Salwan and Sharma 2018). Table 8.1 shows the list of various Streptomyces strains characterized for industrially important enzymes.

To meet the growing demand for industrial enzymes, there is a prerequisite for interminable research and innovations to discovering novel and highly competent enzymes with cost-effective and environmentally friendly production. In addition, the huge accessibility of genomic and proteomic knowledge points to the high throughput manufacturing of high-value enzymes from Streptomyces strains. Abundant genes responsible for the chemical breakdown of the polysaccharides have been identified in Streptomyces genome concluded the microbes are excellent producers of glycosyl or glycoside hydrolases (GH) which decays the composite polysaccharides which are vital to the carbon cycling of environment and also for animal nutrition (Talamantes et al. 2016).

Table 8.1 List of some important industrial enzymes isolated and characterized from Streptomyces strains

Enzymes	Streptomyces strains	References
α-Amylase	*Streptomyces* sp. D1	Chakraborty et al. (2009)
B-Amylase	*S. Gulbargensis* DAS	Syed et al. (2009)
Xylanases	*Streptomyces* spp.SKK1–8 *S. Viridosporus* T7A	Meryandini et al. (2006) Alberton et al. (2009)
α-L-Arabinofuranosidases	*S.lividans* *Streptomyces* sp. SWU10	Vincent et al. (1997) Phuengmaung et al. (2018)
Laccases	*S. ipomoeae* CECT 3341	Blánquez et al. (2017)
α-Glucuronidases	*S. pristinaespiralis*	Fujimoto et al. (2010)
Ferulic acid esterases	*Streptomyces* S10	Mukherjee et al. (2007)
Acetyl xylanesterases	*S. lividans*	Biely et al. (2003)
Mannanases	*Streptomyces* sp. SirexAA-E	Takasuka et al. (2014)
β-Mannosidases	*Streptomyces* sp. S27	Shi et al. (2011)
α-Galactosidases	*S. coelicolor* A3(2)	Temuujin et al. (2016)
Peroxidases	*S. thermoviolaceus*	Tuncer et al. (2009)
Lignin peroxidases	*S. viridosporus* T7A	Gottschalk et al. (2008)
Cellulase	*S. noboritoensis* *S. viridobrunneus*	Arunachalam et al. (2010) Da Vinha et al. (2011)
Chitinase	*S. rubiginosus*	Jha et al. (2016)
Protease	*S. flavogriseus* HS1	Ghorbel et al. (2014)
Keratinase	*S. gulbargensis*	Syed et al. (2009b)
Lipases	*Streptomyces* sp. TEM 33	Vasconcelos et al. (2018)
Halogenases	*Streptomyces toxytricini*	Zeng and Zhan (2011)
Prenyltransferase	*Streptomyces* sp. SN-593	Takahashi et al. (2010)
Pectinase	*S. lydicus*	Jacob et al. (2008)

8.2.3 Streptomyces: As a Source of β-Glucanases

Currently, the enzyme industries are in search of new microbial strains to produce competent enzymes to achieve the necessities of enzymes in a reliable and eco-friendly manner. Advanced enzyme invention platforms from promising microbes offer more possibility and expose formerly unexploited sources of enzyme. Streptomyces are the chief genus of actinobacterial phylum which is well-characterized and documented as ultimate unlimited origin of different compounds with enormous industrial uses (Sathya and Ushadevi 2014; Mukhtar et al. 2017). Like other actinobacterial members, they are saprophytic, aerobic, and Gram-positive microorganisms with a G + C rich genome. They are spore-producing and can grow in diverse environments, however showing the most abundant in soil, distinguished with a distinct earthy odour by the production of geosmin (Dehnad et al. 2010). The Streptomyces genus are well recognized as the largest reservoir of natural bioactive compounds (Terkina et al. 2006). Currently, 75% of recognized natural antibiotics are produced by *Streptomyces* species (Khattab et al. 2016).

Consistent with previous information, *Streptomyces* secretes a variability of carbohydrases, capable to break down numerous soluble and insoluble polysaccharides (Chater et al. 2010). According to Wu et al. (2018), Actinobacteria are considered as better candidates for isolating β-glucan degrading enzymes, however, very little has been explored for potential screening and production of β-glucanase. Some Streptomyces spp. has been recognized as endo-β-1,3-glucanases producers viz. S. rimosus (Beyer and Diekmann 1984), Streptomyces sp. Mo (Kurakake et al. 2013), S. torulosus (Park et al. 2012), S. sioyaensis (Hong et al. 2002) and S. matensis (Woo et al. 2014). The exo-β-1,4-glucanase, a cellulolytic enzyme activity has been studied in S. coelicolor A(3) (Lee et al. 2018), S. reticuli (Walter and Schrempf 1996) and S. thermocerradoensis I3 (Carolina et al. 2015). Even though Streptomyces are the immense resource of industrial enzymes, practice as a platform of industrial production is comparatively limited due to deliberate mycelial growth, inefficient nutrient consumption and the non-existence of advanced expression systems (Edison and Pradeep 2020b). Hence, for the overproduction of industrially important enzymes, well-characterized efficient heterologous production platforms like Escherichia coli or Bacillus subtilis may currently be used.

8.3 Isolation and Screening of β-Glucanase-Producing Actinomycetes Strains

Screening of potential soil actinomycetes is static at the infant phase because less than one part of soil biodiversity has been explored (Baltz 2007). An important factor considered before isolating microorganisms with the potential application is understanding the biodiversity and environmental features associated with growth. The search for distinctive enzymes from unusual ecological habitats is highly fascinating and have great opportunities that may also point to the developments in high throughput screening programmes. Microbial beta-glucanase has been isolated from a variety of microbes and well-characterized (Velho-Pereira and Kamat 2013). Actinomycetes have been extensively recognized as a source of β-glucan degrading enzymes. Amongst the wide genus, Streptomyces are the most prevalent group of enzyme producers (Wu et al. 2018). Some of them include Streptomyces sioyaensis (Hong et al. 2008), Streptomyces sp. 9X166 (Sakdapetsiri et al. 2016), Streptomyces sp. S27 (Shi et al. 2010), Streptomyces matensis ATCC 23935 (Woo et al. 2014), Streptomyces rochei (Wu et al. 2002) and Streptomyces sp. EF-14 (Fayad et al. 2001).

Screening of actinomycetes for beta-glucanase activity is evaluated by measuring semi-quantitative production (primary screening) and quantitative production (secondary screening) of enzymes. Semi-quantitative production is measured by the plate assay method, in which the activity is analyzed based on the clear halo formation around the isolated colonies which directly indicated the region of enzyme action to produce glucose units. The zone of clearance is due to the hydrolysis of

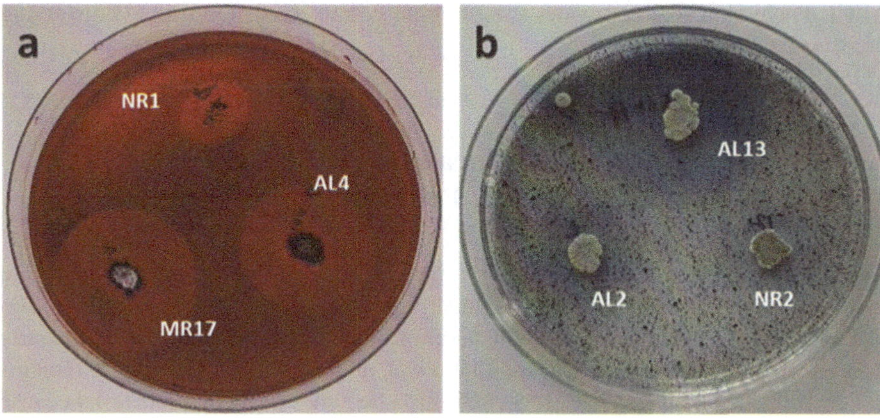

Fig. 8.2 Zone of hydrolysis produced by actinomycetes strains on (**a**) Avicel agar plate and (**b**) AZCL-Pachyman agar plate

substrate by the action of enzymes, which can be detected by staining or by using dye cross-linked substrates. Azurine cross-linked (AZCL) polysaccharide substrates are widely used for screening glycosyl hydrolases (Li et al. 2011; Nyyssönen et al. 2013). Then enzyme activities are calculated by determining the enzymatic index (EI) of each strain by measuring the diameter of the colony and hydrolysis zone. The quantitative approach, however, is regarded as the study's ultimate form of validation since it provides more accurate data with little variability and better sensitivity, allowing comparisons of the relative abundance of enzymes (Bisswanger 2014; Farris et al. 2016). Secondary or quantitative enzyme screening is performed by submerged fermentation followed by the estimation of reducing sugars derived from the hydrolysis of substrates. The actinomycetes strains that produced the highest enzyme activities will be considered as potent enzyme producers and selected for further studies.

Actinomycetes strains isolated from soil samples were screened for exo-β-1,4-glucanase and endo-β-1,3-glucanase activities. The presence of a pale halo around the colonies after Congo red dye staining indicated the production of the enzyme exo-β-1,4-glucanase (Fig. 8.2a). The Congo red dye remains attached to areas where the presence of β-1,4-D-glucanohydrolase bonds (Lamb and Loy 2005). Digestion of substrate Avicel by exo-β-1,4-glucanase produces glucose residues, it does not have any affinity to Congo red. The endo-β-1,3-glucanase activity of isolated actinomycetes strains was screened using 0.2% AZCL-Pachyman as a carbon source in a plate assay. AZCL-Pachyman is an insoluble blue-coloured azurine dye cross-linked substrate specifically for the determination of endo-β-1,3-glucanase activity. The endo-β-1,3-glucananolytic activity of actinomycetes strains produced a blue halo around the colonies indicating the zone of hydrolysis (Fig. 8.2b). This is due to the activity of the enzyme on insoluble

AZCL-Pachyman released water-soluble blue colour dyed fragments (Edison and Pradeep 2020a).

8.4 Identification of Beta-Glucanase-Producing Actinomycetes Strains: A Polyphasic Approach

Actinobacteria is documented as a major lineage in the Bacteria domain. The genera within the phylum exhibit massive diversity in terms of morphological, physiological and metabolic competencies. The taxonomy of Actinobacteria has significantly developed gradually with the accumulation of understandings. Though it remains rather confused and the species-level definition becomes an incomplete task because of the variability in cultural, morphological, physiological, and biochemical characteristics observed at both inter- and intra-species levels (Anderson and Wellington 2001). Consequently, the polyphasic taxonomic approach is more advisable in actinomycetes taxonomy. Molecular genomics, cultural, biochemical, physiological, and chemotaxonomic studies collectively constitute the taxonomic position of microorganisms is named as 'polyphasic taxonomic approach' (Vandamme et al. 1996), which helps to determine the species-level taxonomy within the genus (Anderson and Wellington 2001). Currently, this approach is more popular in bacterial classification and provides new insights into the organism's species and genus level sortings. Accurate taxonomical identification is obligatory for every novel microbial isolate from unusual environmental habitats. The key characteristics used to define the taxonomy of Actinobacteria into genus and species levels includes morphological studies based on microscopy, cellular chemistry-based chemotaxonomy, cultural studies mainly on growth mycelia formation and pigments, biochemical-physiological characteristics and finally molecular genotypic studies based on 16 s rRNA gene sequencing (Acharyabhatta et al. 2013; Barka et al. 2016).

In actinobacteria, Streptomyces spp. are mostly exploited for the screening of beta-glucanolytic enzymes. Waksman and Henrici (1943) was proposed the genus *Streptomyces* and categorized it in Streptomycetaceae family based on morphology and cell wall chemotype. The progress of numerical classification employed only phenotypic traits resolved the intergeneric relationship within the family, but not disentangled the taxonomy within the genus. The genus Streptomyces are the producers of beyond half of the currently documented bioactive compounds in the world and offered pronounced industrial and academic interest. Despite this, their taxonomy leftovers rather confused and the species definition is unresolved because of the diversity of morphological, cultural, physiological, and biochemical characteristics observed at the intra- and the inter-species levels. The Difficulties in classifying new strains highlight the need for a polyphasic approach in Streptomyces taxonomy which employs a combination of genotypic, phenotypic, and phylogenetic tests (Anderson and Wellington 2001).

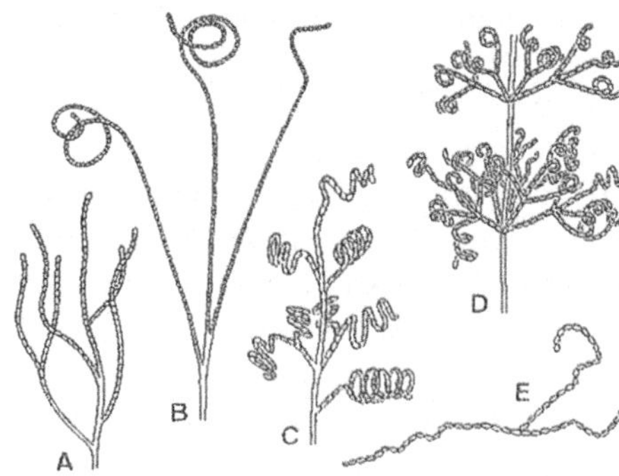

Fig. 8.3 Streptomyces Spore chain configuration. (**a**) Rectiflexibiles type, (**b**) Retinaculiaperti type, (**c**) Spira type, (**d**) Verticillati type, (**e**) fragmenting branched aerial hyphae (Li et al. 2016)

The International Streptomyces Project (ISP) in 1964 introduced by standard criteria for the species determination of Streptomyces, where, Shirling and Gottlieb (1968a, b, 1969, 1972) described the use of aerial mycelium morphology, the colour of aerial and substrate mycelium, production of soluble and melanin pigment and the utilization of certain carbon sources. The ISP described 450 species of Streptomyces. Based on the ISP data numerous classification keys have been described (Kuster 1972; Nonomura 1974; Szabo et al. 1975). The phenotypic characteristics such as the aerial mycelium morphology comprise the mode of branching and spore chain configuration (Fig. 8.3) and spore surface is regarded as significant for Streptomyces classification (Pridham et al. 1958). A numerical taxonomic approach was described by Williams et al. (1983) which studied cell wall chemotype using phenotypic traits. Streptomyces differs from other actinomycetes in its cell wall properties, which include the lack of unique sugars, the presence of LL-diaminopimelic acid and glycine and acetyl muramyl residues in peptidoglycans (Uchida and Seino 1997). A chemotaxonomic approach includes fatty acid profiling, which has long been practised to distinguish Streptomyces species (Saddler et al. 1987). Goodfellow (1988) described some biochemical tests for Streptomyces taxonomy. The genome analysis through molecular techniques has contributed significant knowledge of Streptomyces taxonomy, comprises rRNA gene sequencing and evaluation, which helps the taxonomical identification from genus, species and extended up to strain level (Hain et al. 1997). The polyphasic approach comprises the usefulness of all these methods in the taxonomic evaluation of natural Streptomyces isolates.

At present, numerous novel Streptomyces species and strains are described based on a polyphasic taxonomic approach. The cumulative result of polyphasic characterization elucidate the taxonomic position of efficient β-glucanase producing Western Ghats actinomycetes isolates TBG-MR17 and TBG-AL13 as Streptomyces althioticus TBG-MR17 and Streptomyces cinereoruber sub sp. cinereoruber TBG-AL13, respectively (Edison and Pradeep 2020a).

Acharyabhatta et al. (2013) isolated alkaline amylase-producing Streptomyces rochei BTSS 1001 from marine sediments of the southeast coast of Bay of Bengal and studied taxonomy through a polyphasic approach. Tanasupawat et al. (2016) described a novel species, Streptomyces actinomycinicus isolated from peat swamp forest soil of Rayong Province, Thailand using a polyphasic approach. Van der Aart et al. (2018) used polyphasic taxonomy for describing the classification of novel strain Streptomyces roseofaciens isolated from QinLing Mountains and detected more than 40 metabolomic gene clusters, codes natural products. Charousová et al. (2018) determined the taxonomic position of most active antibiotic producers, Streptomyces globosus DK15 and Streptomyces ederensis ST13 strains using a polyphasic approach. Landwehr et al. (2018) used a polyphasic approach for clarifying the taxonomic position of the novel roseoflavin-producing isolate Streptomyces davaonensis sp. nov. JCM 4913 T and cinnabaramide-producing isolate as Streptomyces cinnabarigriseus sp. nov. JS360T.

8.5 The Western Ghats in Kerala: Look for Glucanolytic Streptomyces spp.

The Western Ghats in India is a distinguished biodiversity hot spot of rich flora and fauna with highly productive ecosystems. The area covers signify geomorphic features of enormous importance with a unique ecological niche, exclusively with old mountain ranges starting from Central Maharashtra and stretching up to the extreme Southern point of Kerala. It is a tremendously distinguished biodiversity hot spot of gorgeous flora and fauna and is recognized by UNESCO as a World Heritage Site (Ruckmani and Chakrabarti 2011). Moreover, because of the dynamic rich biodiversity, nearly all regions are enduring unexplored, uninterrupted and are massively treasured with potential microorganisms. Hence, the Western Ghats areas are also mentioned as microbial biodiversity treasure houses. The soil samples were immensely utilized for the isolation of industrially important novel enzyme-producing microorganisms (Nampoothiri et al. 2013). Forest soils are the huge domicile of taxonomically diverse actinomycetes strains, especially Streptomyces sp. Despite the presence of rich recalcitrant biopolymers, they are actively involved in forest nutrient turnover (Bontemps et al. 2013). The exploitation of unique, natural, highly documented and less explored biodiversity ecosystems for actinomycetes is highly necessary for the discovery of novel bioactive metabolites with prospective applications (Jalaja et al. 2011; Mohandas et al. 2012; Balachandran et al. 2012; Nampoothiri et al. 2013).

The search for isolating novel and effective microorganisms from unexplored habitats around the world continues to be an important sector of research. Microbial potential studies focused on extreme environments can be exploited to produce innovations in enzyme technology that may become the future preface of green biotechnology (Edison and Pradeep 2020c). The uniqueness in environmental conditions certainly affected the Streptomyces distribution (Mitra et al. 2008; Jagan Mohan et al. 2013). As a result of extensive screening, Streptomyces strains from these natural and well-treasured habitats have been commercially exploited (Arasu et al. 2008; Balachandran et al. 2015; Rao et al. 2015).

The Western Ghats is one of the treasurable natural resources of the earth (Gadgil 1979). Mostly it has protected and conserved unique biodiversity, pointing out the chance to discover unidentified microorganisms. The β-glucanase-producing Streptomyces spp. are less characterized from the Western Ghats. These regions remain less explored, and also due to inextricably altered habitat the chances of obtaining potential strains of actinomycetes including Streptomyces species with exceptional β-glucanase activities are much higher. Soil samples from Western Ghats areas showed very negligible variations in pH, ranging from 6.20 to 7.50 and electrical conductivity (EC) was laid between 0.70 and 3.30 ds.m-1 (Edison et al. 2017). pH and EC are the most significant parameters for measuring soil quality, directly related to soil microbial biomass (Mohd-Aizat et al. 2014). The weak acidity of Western Ghats soil is due to the accrual of organic acids by the decomposition of surface high leaf litter (Chen and Yang 2000). Soil pH greatly influences the load and diversity of actinomycetes strains. The highest numbers of actinomycetes were obtained on and above pH 7.0. EC indicates the accumulation of soluble salts in soils. Actinobacterial population and metabolic activity may be affected by the high EC of soil. Lower EC of Western Ghats soil also promotes higher actinobacterial load with efficient enzymatic activities. Especially fast-growing prerequisites for enzymes in diverse extents demands an urgent need to explore actinomycetes as a treasured source of marketable enzymes. Western Ghats ecosystems are unusual habitats for promising actinobacterial diversity with extraordinary β-glucanase activity. An extensive range of climatic environments, rich woodland areas and fewer discrepancies in soil type with less acidic to alkaline pH and low EC are relatively favourable for the largest distribution of actinobacteria with high β-glucanase activity. Pre-treatment of collected soil samples with 1% $CaCO_3$ (1%) helps to produce better actinobacterial colony counts. $CaCO_3$ alters the pH of the soil, thus favours the growth of actinobacterial spores and stimulates the formation of aerial mycelia (Natsume et al. 1989). It is one of the good methods for enriching actinomycetes propagules, which significantly yielded high total plate counting. This method also proved effective in increasing the number of rare isolates.

A self-study conducted in Western Ghats regions in Kerala for the screening of beta-glucanases-producing actinomycetes strains resulted in a total of 127 obtained from 10 different areas. A qualitative enzyme assay revealed that only 79 strains (62%) produced endo-β-1,3-glucanase activity, whereas 106 strains (83%) showed exo-β-1,4-glucanase enzyme activity. According to quantitative activity profiling, the actinomycetes strains TBG-MR17 (Fig. 8.4) and TBG-AL13 (Fig. 8.5) are recognized as respective dominant exo-β-1,4-glucanase and endo-β-1,3-glucanase producers. Under optimum culture conditions actinomycetes strains, Streptomyces althioticus TBG-MR17 produced 141 $U.mL^{-1}$ exo-β-1,4-glucanase and Streptomyces cinereoruber sub sp. cinereoruber TBG-AL13 produced 892 U.mL-1 endo-β-1,3-glucanase. This is the first report of the exploration of Western Ghats actinomycetes for both exo-β-1,4-glucanase and endo-β-1,3-glucanase enzymes with tremendously high activities (Edison and Pradeep 2020a).

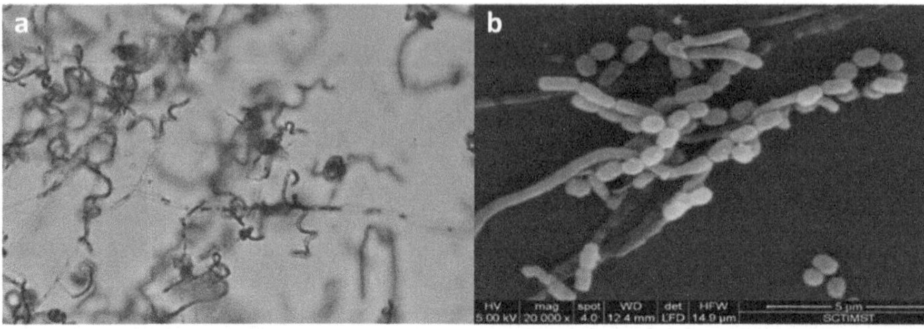

Fig. 8.4 Exo-β-1,4-glucanase producing Streptomyces althioticus TBG-MR17. (**a**) The light microscopic view showed spirals (S) spore chains. (**b**) Scanning electron microscopic view showed oval-rod-shaped spores with spiny surface topography

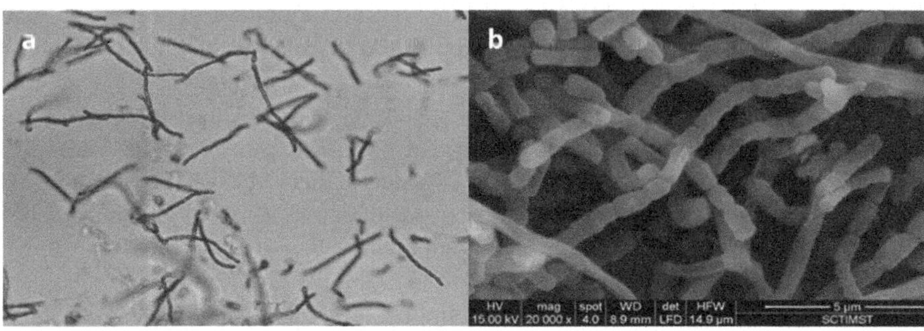

Fig. 8.5 Endo-β-1,3-glucanase-producing Streptomyces cinereoruber sub sp. cinereoruber TBG-AL13. (**a**) The light microscopic view showed the stain has Rectiflexibiles (RF) spore chains. (**b**) Scanning electron microscopic view showed oblong spores with a smooth surface

References

Acharyabhatta A, Kandula SK, Terli R (2013) Taxonomy and polyphasic characterization of alkaline amylase producing marine actinomycete Streptomyces rochei BTSS 1001. Int J Microbiol 276921:1–8

Aderem A (2005) Systems biology: its practice and challenges. Cell 121(4):511–513

Alberton LR, Vandenberghe LPS, Assmann R, Fendrich RC, Rodriguéz-León J, Soccol CR (2009) Xylanase production by Streptomyces viridosporus T7A in submerged and solid-state fermentation using agro-industrial residues. Braz Arch Biol Technol 52(spe):171–180

Anderson AS, Wellington EMH (2001) The taxonomy of Streptomyces and related genera. Int J Syst Evol Microbiol 51(3):797–814

Arasu MV, Duraipandiyan V, Agastian P, Ignacimuthu S (2008) Antimicrobial activity of Streptomyces spp. ERI-26 recovered from Western Ghats of Tamil nadu. J Mycol Med 18(3): 147–153

Arunachalam R, Wesely EG, George J, Gurusamy A (2010) Novel approaches for identification of Streptomyces noboritoensis TBG-V20 with cellulase production. Curr Res Bacteriol 3:15–26

Balachandran C, Duraipandiyan V, Emi N, Ignacimuthu S (2015) Antimicrobial and cytotoxic properties of Streptomyces sp. (ERINLG-51) isolated from southern Western Ghats. sijbs 1(1): 7–14
Balachandran C, Duraipandiyan V, Ignacimuthu S (2012) Cytotoxic (A549) and antimicrobial effects of Methylobacterium sp. isolate (ERI-135) from Nilgiris forest soil. India Asian Pac J Trop Biomed 2(9):712–716
Baltz HR (2007) Antimicrobials from Actinomycetes: Back to the future. Microbe 2:125
Barka EA, Vatsa P, Sanchez L, Gaveau-Vaillant N, Jacquard C, Meier-Kolthoff JP, Klenk HP, Clément C, Ouhdouch Y et al (2016) Taxonomy, physiology, and natural products of Actinobacteria. Microbiol Mol Biol Rev 80(1):1–43
Beyer M, Diekmann H (1984) The laminarinase system of Streptomyces sp. ATCC 11238. Appl Microbiol Biotechnol 20(3):207–212
Bhattacharyya B (1998) Pal, S, & Sen, S. Anitibiotic prod Streptomyces Hygroscopicus D1.5. Cultural effect, Revista de Microbiologia 29(3):254–257
Biely P, Mastihubova M, Cete GL, Richard VG (2003) Mode of action of acetylxylan esterase from Streptomyces lividans: a study with deoxy and deoxy-fluoro analogues of acetylated methyl β-d-xylopyranoside. Biochim Biophys Acta (BBA) Gen Subj 2:82–88
Bisswanger H (2014) Enzyme assays Perspect Sci 1(1–6):41–55
Blánquez A, Ball AS, González-Pérez JA, Jiménez-Morillo NT, González-Vila F, Arias ME, Hernández M (2017) Laccase SilA from Streptomyces ipomoeae CECT 3341, a key enzyme for the degradation of lignin from agricultural residues? PLoS One 12(11):e0187649
Bontemps C, Toussaint M, Revol PV, Hotel L, Jeanbille M, Uroz S, Turpault MP, Blaudez D, Leblond P (2013) Taxonomic and functional diversity of Streptomyces in a forest soil. FEMS Microbiol Lett 342(2):157–167
Carolina CQ, Cunha B, Gama AR, Jesuino RSA, Faria FP, Bataus LAM (2015) Production of cellulases from a novel thermophilic Streptomyces thermocerradoensis I3 using agricultural waste residue as substrate. J Agric Environ Sci 4(1):90–99
Chakraborty S, Khopade A, Kokare C, Mahadik K, Chopade B (2009) Isolation and characterization of novel α-amylase from marine Streptomyces sp. J Mol Cat B Enzym 58(1–4):17–23
Champness WC (2000) Actinomycete development, antibiotic production and phylogeny: questions and challenges. In: Brun YV, Skimkets LJ (eds) Prokaryotic development: Washington. American Society for Microbiology, DC, pp 11–31
Charousová I, Medo J, Hleba L, Javoreková S (2018) Streptomyces globosus DK15 and Streptomyces ederensis ST13 as new producers of factumycin and tetrangomycin antibiotics. Braz J Microbiol 49(4):816–822
Chater KF, Biró S, Lee KJ, Palmer T, Schrempf H (2010) The complex extracellular biology of Streptomyces. FEMS Microbiol Rev 34(2):171–198
Chater KF (1993) Genetics of differentiation in Streptomyces. Annu Rev Microbiol 47:685–713
Chen WS, Yang SS (2000) Organic acid contents in Tatachoest soil. J Exp For 14:99–108
Colquhoun JA, Mexson J, Goodfellow M, Ward AC, Horikoshi K, Bull AT (1998) Novel rhodococci and other mycolate actinomycetes from the deep sea. Antonie Leeuwenhoek 74(1–3):27–40
Da Vinha FN, Gravina-Oliveira MP, Franco MN, Macrae A, da Silva Bon EP, Nascimento RP, Coelho RR (2011) Cellulase production by Streptomyces viridobrunneus SCPE-09 using lignocellulosic biomass as inducer substrate. Appl Biochem Biotechnol 164(3):256–267
Dehnad A, Parsa Yeganeh L, Bakhshi R, Mokhtarzadeh A (2010) Investigation antibacterial activity of Streptomycetes isolates from soil samples west of Iran. Afr J Microbiol Res 4(16): 1685–1693
Diraviyam T, Radhakrishnan M, Balagurunathan R (2011) Antioxidant activity of melanin pigment from Streptomyces species D5 isolated from desert soil, Rajasthan, India. Drug Invent Today 3: 12–13

Edison LK, Anilkumar R, Gayathri V, Pradeep NS (2017) Enzyme activity profiling of exo-β-1,4-glucanase, endo-β-1,3-glucanase and protease in Streptomyces species from highland, midland and lowland areas of Kerala. India Int J Adv Res 5(2):2431–2444

Edison LK, Pradeep NS (2020a) Functional screening of β-glucanase producing Actinomycetes strains from Western Ghats ecosystems of Kerala. India Indian J Appl Pure Biogr 35(2): 255–270

Edison LK, Pradeep NS (2020b) Beta-Glucanolytic soil Actinomycetes: diversity and applications. In: Sharma SK, Singh UB, Sahu PK, Singh HV, Sharma PK (eds) Rhizosphere microbes. Microorganisms for sustainability, vol 23. Springer, Singapore, pp 471–482. https://doi.org/10.1007/978-981-15-9154-9_19

Edison LK, Pradeep NS (2020c) Actinobacterial nanoparticles: green synthesis, evaluation and applications. In: Patra J, Fraceto L, Das G, Campos E (eds) Green nanoparticles. Nanotechnology in the life sciences. Springer, Cham, pp 371–384. https://doi.org/10.1007/978-3-030-39246-8_20

Farris, M. H., Ford, K. A., & Doyle, R. C. (2016). Qualitative and quantitative assays for detection and characterization of protein antimicrobials. Journal of Visualized Experiments, 10(110), e53819.

Fayad KP, Simao-Beaunoir AM, Gauthier A, Leclerc C, Mamady H, Beaulieu C, Brzezinski R (2001) Purification and properties of a beta-1,6-glucanase from Streptomyces sp. EF-14, an actinomycete antagonistic to Phytophthora spp. Appl Microbiol Biotechnol 57(1–2):117–123

Filippova SN, Vinogradova KA (2017) Programmed cell death as one of the stages of streptomycete differentiation. Microbiology 86(4):439–454

Fujimoto Z, Ichinose H, Biely P, Kaneko S (2010) Crystallization and preliminary crystallographic analysis of the glycoside hydrolase family 115 α-glucuronidase from Streptomyces pristinaespiralis. Acta Crystallogr F 67(1):68–71

Gadgil M (1979) Hills, dams and forests. Some field observations from the Western Ghats. Proceedings of the Indian Academy of Science 2(3):291–303

Ghorbel S, Kammoun M, Soltana H, Nasri M, Hmidet N (2014) Streptomyces flavogriseus HS1: isolation and characterization of extracellular proteases and their compatibility with laundry detergents. Biomed Res Int 345980:1–8

Goodfellow M (1988) Numerical taxonomy and selective isolation of industrially important actinomycetes. Actinomycetologica 2(1):13–29

Gottschalk LMF, Bon EP, Nobrega R (2008) Lignin peroxidase from Streptomyces viridosporus T7 a: enzyme concentration using ultrafiltration. Appl Biochem Biotechnol 147(1–3):23–32

Hain T, Ward-Rainey N, Kroppenstedt RM, Stackebrandt E, Rainey FA (1997) Discrimination of Streptomyces albidoflavus strains based on the size and number of 16S-23S ribosomal DNA intergenic spacers. Int J Syst Bacteriol 47(1):202–206

Hasani A, Kariminik A, Issazadeh K (2014) Streptomycetes: characteristics and their antimicrobial activities. Int J Adv Biol Biomed Res 2(1):63–75

Hong TY, Cheng CW, Huang JW, Meng M (2002) Isolation and biochemical characterization of an endo-1,3-β-glucanase from Streptomyces sioyaensis containing a C-terminal family 6 carbohydrate-binding module that binds to 1,3-β-glucan. Microbiology (Reading) 148(4):1151–1159

Hong TY, Hsiao YY, Meng M, Li TT (2008) The 1.5 a structure of endo-1,3-beta-glucanase from Streptomyces sioyaensis: evolution of the active-site structure for 1,3-beta-glucan-binding specificity and hydrolysis. Acta Crystallogr D Biol Crystallogr 64(9):964–970

Hopwood DA (2007) Streptomyces in nature and medicine: the antibiotic makers. Oxford University Press, New York, p 250

Jacob N, Asha Poorna CA, Prema P (2008) Purification and partial characterization of polygalacturonase from Streptomyces lydicus. Bioresour Technol 99(14):6697–6701

Jagan Mohan YSYV, Sirisha B, Haritha R, Ramana T (2013) Selective screening isolation and characterization of antimicrobial agents from marine actinomycetes. J Pharm Sci 5:443–449

Jalaja V, Swetha S, Kumar SS, Deepthi A, Kumar RS, Pandey A (2011) Isolation and characterization of alpha amylase from a metagenomic library of Western Ghats of Kerala, India. Biologia 66:939–944

Jha S, Modi HA, Jha CK, Yildiz F (2016) Characterization of extracellular chitinase produced from Streptomyces rubiginosus isolated from rhizosphere of Gossypium sp. Cogent Food Agric 2(1): 1–12

Khattab AI, Babiker EH, Saeed HA (2016) Streptomyces: isolation, optimization of culture conditions and extraction of secondary metabolites. Int Curr Pharm J 5(3):27–32

Kurakake M, Yamanouchi Y, Kinohara K, Moriyama S (2013) Enzymatic properties of β-1,3-glucanase from Streptomyces sp Mo. J Food Sci 78(4):C502–C506

Küster E, Williams ST (1964) Selection of media for isolation of Streptomycetes. Nature 202(4935):928–929

Kuster E (1968) Taxonomy of soil actinomycetes and related organisms. In: Gray S, Parkinson T (eds) Ecology of soil bacteria. Liverpool University Press, Liverpool, pp 322–336

Kuster E (1972) Simple working key for the classification and identification of named taxa included in the international Streptomyces project. Int J Syst Bacteriol 22(3):139–148

Lamb J, Loy T (2005) Seeing red: the use of Congo red dye to identify cooked and damaged starch grains in archaeological residues. J Archaeol Sci 32(10):1433–1440

Landwehr W, Kämpfer P, Glaeser SP, Rückert C, Kalinowski J, Blom J, Goesmann A, Mack M, Schumann P et al (2018) Taxonomic analyses of members of the Streptomyces cinnabarinus cluster, description of Streptomyces cinnabarigriseus sp nov and Streptomyces davaonensis sp nov. Int J Syst Evol Microbiol 68(1):382–393

Lee CR, Chi WJ, Lim JH, Dhakshnamoorthy V, Hong SK (2018) Expression and characterization of the processive exo-β-1,4-cellobiohydrolase SCO6546 from Streptomyces coelicolor a(3). J Basic Microbiol 58(4):310–321

Li LL, Taghavi S, McCorkle SM, Zhang YB, Blewitt MG, Brunecky R, Adney WS, Himmel ME, Brumm P et al (2011) Bioprospecting metagenomics of decaying wood: mining for new glycoside hydrolases. Biotechnol Biofuels 4(1):23

Li Q, Chen X, Jiang Y, Jiang C (2016) Morphological identification of Actinobacteria. In: Dhanasekaran D, Jiang Y (eds) Actinobacteria–basics and biotechnological applications. Tech, Rijeka, pp 59–86

Ludwig W, Euzeby J, Schumann P, Buss HJ, Trujillo ME, Kampfer P, Whiteman WB (2012) Road map of the phylum Actinobacteria. In: Goodfellow M et al.. Bergey's manual of systematic bacteriology vol. 5: New York: Springer-Verlag, pp. 1–28

Macagnan D, Romeiro RS, de Souza JT, Pomella AWV (2006) Isolation of actinomycetes and endospore-forming bacteria from the cacao pod surface and their antagonistic activity against the witches' broom and black pod pathogens. Phytoparasitica 34(2):122–132

Manteca A, Alvarez R, Salazar N, Yagüe P, Sanchez J (2008) Mycelium differentiation and antibiotic production in submerged cultures of Streptomyces coelicolor. Appl Environ Microbiol 74(12):3877–3886

Manteca A, Sanchez J (2010) Streptomyces developmental cycle and secondary metabolite production. Curr Res Technol Educ Top Appl Microbiol Microb Biotechnol 1:560–566

Meryandini A, Hendarwin T, Saprudin D, Lestari Y (2006) Characterization of xylanase Streptomyces spp. SKK1-8. HAYATI. J Biosci 13(4):151–155

Mitra A, Santra SC, Mukherjee J (2008) Distribution of actinomycetes, their antagonistic behaviour and the physico chemical characteristics of the world's largest tidal mangrove forest. Appl Microbiol Biotechnol 80(4):685–695

Mohandas SP, Ravikumar S, Menachery SJ, Suseelan G, Narayanan SS, Nandanwar H, Nampoothiri KM (2012) Bioactives of microbes isolated from Western Ghats belt of Kerala show lactamase inhibition along with wide spectrum antimicrobial activity. Appl Biochem Biotechnol 167(6):1753–1762

Mohanraj G, Sekar T (2013) Isolation and screening of actinomycetes from marine sediments for their potential to produce antimicrobials. Int J Life Sci Pharm Res 2(3):115–126

Mohd-Aizat A, Mohamad-Roslan MK, Sulaiman WNA, Karam DS (2014) The relationship between soil pH and selected soil properties in 48 years logged-over forest. Int J Environ Sci 4(6):1129–1140

Moncheva P, Tishkov S, Dimitrova N, Chipeva V, Bogatzevska N (2002) Characteristics of soil actinomycetes from Antarctica. J Cult Coll 3:3–14

Mukherjee G, Singh RK, Mitra A, Sen SK (2007) Ferulic acid esterase production by Streptomyces sp. Bioresour Technol 98(1):211–213

Mukhtar S, Zaheer A, Aiysha D, Abdulla Malik KA, Mehnaz S (2017) Actinomycetes: a source of industrially important enzymes. J Proteomics Bioinform 10(12):316–319

Nampoothiri KM, Ramkumar B, Pandey A (2013) Western Ghats of India: rich source of microbial diversity. J Sci Ind Res 72:617–623

Natsume M, Yasui K, Marumo S (1989) Calcium ion regulates aerial mycelium formation in actinomycetes. J Antibiot (Tokyo) 42(3):440–447

Nonomura H (1974) Key for classification and identification of 458 species of the Streptomyces included in ISP. J Ferment Technol 52:78–92

Nyyssönen M, Tran HM, Karaoz U, Weihe C, Hadi MZ, Martiny JB, Martiny AC, Brodie EL (2013) Front Microbiol 4:282

Pamboukian CRD, Guimarães LM, Facciotti MCR (2002) Applications of image analysis in the characterization of Streptomyces olindensis in submerged culture. Braz J Microbiol 33(1): 17–21

Park JK, Kim J, Park YI, Kim S (2012) Purification and characterization of a 1,3-β-D-glucanase from Streptomyces torulosus PCPOK-0324. Carbohydr Polym 87(2):1641–1648

Pathom-Aree W, Stach JE, Ward AC, Horikoshi K, Bull AT, Goodfellow M (2006) Diversity of actinomycetes isolated from challenger deep sediment (10,898 m) from the Mariana trench. Extremophiles 10(3):181–189

Phuengmaung P, Kunishige Y, Sukhumsirichart W, Sakamoto T (2018) Identification and characterization of GH62 bacterial α-L-arabinofuranosidase from thermotolerant Streptomyces sp. SWU10 that preferentially degrades branched L-arabinofuranoses in wheat arabinoxylan. Enzym Microb Technol 112:22–28

Pridham TG, Hesseltine CW, Benedict RG (1958) A guide for the classification of Streptomycetes according to selected groups. Placement of strains in morphological sections. Appl Microbiol 6(1):52–79

Rao HCY, Rakshith D, Satish S (2015) Antimicrobial properties of endophytic actinomycetes isolated from Combretum latifolium Blume, a medicinal shrub from Western Ghats of India. Front Biol 10(6):528–536

Ruckmani A, Chakrabarti T (2011) Analysis of bacterial community composition of a spring water from the Western Ghats, India using culture dependent and molecular approaches. Curr Microbiol 62(1):7–15

Saddler GS, O'Donnell AG, Goodfellow M, Minnikin DE (1987) SIMCA pattern recognition in the analysis of streptomycete fatty acids. Microbiology 133(5):1137–1147

Sakdapetsiri C, Fukuta Y, Aramsirirujiwet Y, Shirasaka N, Kitpreechavanich V (2016) Antagonistic activity of endo-β-1,3-glucanase from a novel isolate, Streptomyces sp. 9X166, against black rot in orchids. J Basic Microbiol 56(5):469–479

Salwan R, Sharma V (2018) The role of Actinobacteria in the production of industrial enzymes. In: Singh BP et al (eds) New and future developments in microbial biotechnology and bioengineering. Elsevier, pp 165–177

Sathya R, Ushadevi T (2014) Industrially important enzymes producing Streptomyces species from mangrove sediments. Int J Pharm Pharm Sci 6(10):233–237

Shi P, Yao G, Cao Y, Yang P, Yuan T, Huang H, Bai Y, Yao B (2011) Cloning and characterization of a new β-mannosidase from Streptomyces sp. S27. Enzym Microb Technol 49(3):277–283

Shi P, Yao G, Yang P, Li N, Luo H, Bai Y, Wang Y, Yao B (2010) Cloning, characterization, and antifungal activity of an endo-1,3-beta-D: -glucanase from Streptomyces sp. S27. Appl Microbiol Biotechnol 85(5):1483–1490

Shirling EB, Gottlieb D (1968a) Cooperative description of type cultures of Streptomyces. II. Species descriptions from first study. Int J Syst Bacteriol 18(2):69–189
Shirling EB, Gottlieb D (1968b) Cooperative description of type cultures of Streptomyces III. Additional species descriptions from first and second studies. Int J Syst Bacteriol 18(4):279–392
Shirling EB, Gottlieb D (1969) Cooperative description of type cultures of Streptomyces IV. Species descriptions from the second, third and fourth studies. Int J Syst Bacteriol 19(4): 391–512
Shirling EB, Gottlieb D (1972) Cooperative description of type cultures of Streptomyces IV. Additional descriptions. Int J Syst Bacteriol 22(4):265–394
Srinivasan MC, Laxman RS, Deshpande MV (1991) Physiology and nutrition aspects of actinomycetes–an overview. World J Microbiol Biotechnol 7(2):171–184
Srivibool R, Kurakami K, Sukchotiratana M, Tokuyama S (2004) Coastal soil actinomycetes: Thermotolerant strains producing N-Acylamino acid racemase. Sci Asia 30(2):123–126
Subbarao N (1999) Soil microbiology, 4th edn. Science Publishers, Inc, USA, pp 279–283
Syed DG, Agasar D, Pandey A (2009) Production and partial purification of β-amylase from a novel isolate Streptomyces gulbargensis. J Ind Microbiol Biotechnol 36(2):189–194
Syed DG, Lee JC, Li WJ, Kim CJ, Agasar D (2009b) Production, characterization and application of keratinase from Streptomyces gulbargensis. Bioresour Technol 100(5):1868–1871
Szabo I, Marton M, Buti I, Fernandez C (1975) A diagnostic key for the identification of species of Streptomyces and Streptoverticillium included in the international Streptomyces project. Acta Bot Acad Sci Hung 21:387–418
Takahashi S, Takagi H, Toyoda A, Uramoto M, Nogawa T, Ueki M, Sakaki Y, Osada H (2010) Biochemical characterization of a novel indole prenyltransferase from Streptomyces sp. SN-593. J Bacteriol 192(11):2839–2851
Takahashi Y, Omura S (2003) Isolation of new actinomycete strains for the screening of new bioactive compounds. J Gen Appl Microbiol 49(3):141–154
Takasuka TE, Acheson JF, Bianchetti CM, Prom BM, Bergeman LF, Book AJ, Currie CR, Fox BG (2014) Biochemical properties and atomic resolution structure of a proteolytically processed β-mannanase from cellulolytic Streptomyces sp. SirexAA-E PLoS One 9(4):e94166
Talamantes D, Biabini N, Dang H, Abdoun K, Berlemont R (2016) Natural diversity of cellulases, xylanases, and chitinases in bacteria. Biotechnol Biofuels 9:133
Tanasupawat S, Phongsopitanun W, Suwanborirux K, Ohkuma M, Kudo T (2016) Streptomyces actinomycinicus sp. nov., isolated from soil of a peat swamp forest. Int J Syst Evol Microbiol 66(1):290–295
Temuujin U, Park JS, Hong SK (2016) Molecular characterization of the α-galactosidase SCO0284 from Streptomyces coelicolor A3(2), a family 27 Glycosyl hydrolase. J Microbiol Biotechnol 26(9):1650–1656
Terkina IA, Parfenova VV, Ahn TS (2006) Antagonistic activity of Actinomycetes of Lake Baikal. Appl Biochem Microbiol 42(2):173–176
Tuncer M, Kuru A, Sahin N, Isikli M, Isik K (2009) Production and partial characterization of extracellular peroxidase produced by Streptomyces sp. F6616 isolated in Turkey. Ann Microbiol 59(2):323–334
Uchida K, Seino A (1997) Intra- and intergeneric relationships of various actinomycete strains based on the acyl types of the muramyl residue in cell wall peptidoglycans examined in a glycolate test. Int J Syst Bacteriol 47(1):182–190
Van der Aart L, Nouinoui I, Kloosterman A, Ingual JM, Willemse J, Goodfellow M, van Wezel G (2018) Classification of the gifted natural product producer Streptomyces roseofaciens sp. nov. by polyphasic taxonomy. bioRxiv
Vandamme P, Pot B, Gillis M, de Vos P, Kersters K, Swings J (1996) Polyphasic taxonomy, a consensus approach to bacterial systematics. Microbiol Rev 60(2):407–438
Vasconcelos ES, Salla RF, Cruz-Hernandez IL (2018) Production of lipase from Streptomyces. J Bacteriol Mycol 6(2):102–103

Velho-Pereira S, Kamat NM (2013) Actinobacterial research in India. Indian J Exp Biol 51(8): 573–596

Vincent P, Shareck F, Dupont C, Morosoli R, Kluepfel D (1997) New alpha-L-arabinofuranosidase produced by Streptomyces lividans: cloning and DNA sequence of the abfB gene and characterization of the enzyme. Biochem J 322(3):845–852

Waksman SA, Henrici AT (1943) The nomenclature and classification of the actinomycetes. J Bacteriol 46(4):337–341

Walter S, Schrempf H (1996) The synthesis of the Streptomyces reticuli cellulase (avicelase) is regulated by both activation and repression mechanisms. Mol Gen Genet 251(2):186–195

Willemse J, Borst JW, de Waal E, Bisseling T, van Wezel GP (2011) Positive control of cell division: FtsZ is recruited by SsgB during sporulation of Streptomyces. Genes Dev 25(1):89–99

Williams ST, Goodfellow M, Alderson G, Wellington EMH, Sneath PHA, Sackin MJ (1983) Numerical classification of Streptomyces and related genera. J Gen Microbiol 129(6): 1743–1813

Williams ST, Goodfellow M, Alderson G (1989) Genus Streptomyces. Waksman and Henrici 1943, 339AL. In: Williams ST et al (eds) Bergey's manual of determinative bacteriology, vol 4. Williams & Wilkins, Baltimore, pp 2453–2492

Woo JB, Kang HN, Woo EJ, Lee SB (2014) Molecular cloning and functional characterization of an endo-β-1,3-glucanase from Streptomyces matensis ATCC 23935. Food Chem 148(148): 184–187

Wu H, Shimoi H, Ito K (2002) Purification and characterization of beta-1,6-glucanase of Streptomyces rochei application in the study of yeast cell wall proteins. Biosci Biotechnol Biochem 66(11):2515–2519

Wu Q, Dou X, Wang Q, Guan Z, Cai Y, Liao X (2018) Isolation of β-1,3-Glucanase-producing microorganisms from Poria cocos cultivation soil via molecular biology. Molecules 23(7):1555

Zeng J, Zhan J (2011) Characterization of a tryptophan 6-halogenase from Streptomyces toxytricini. Biotechnol Lett 33(8):1607–1613

Chapter 9
Structural Characterization of Beta-Glucanase from Actinobacteria

Lekshmi K. Edison, T. G. Nandu, and N. S. Pradeep

Abstract Beta-glucanases have prominent applications in the brewing industry as well as the feed enzyme industry. *Streptomyces* origin beta-glucanases have more demand due to the broad substrate specificity, high catalytic activity, and stability in wide ranges of pH and temperatures. Many industrially valuable beta-glucanase are identified from novel actinomycetes strains. One of the strategies for the identification of novel glucanohydrolase includes the prediction of the enzymatic function of the uncharacterized gene sequence using in silico approaches. The gene sequence and corresponding protein structure of a particular enzyme contain all the specificity of the enzyme. The theoretical overview of predicted proteins and their coding genes provides a basic concept of the protein domain architecture that is likely to regulate their function, stability and their interaction with substrates. The in silico study of structure–function relationships exclusively reveals the absolute potential of beta-glucanase enzymes. The insights provided by in silico tools on protein structure and function can be channelized for increased enzyme production and stability towards industrial applications.

9.1 Introduction

Beta-glucanase encoding genes have been extensively explored in different organisms because of their enormous importance in industrial applications. Recently, it has been concerned with commercial attentiveness due to the importance of beer fermentation and as a poultry feed formulation (von Wettstein et al. 2000; Lamp et al. 2015) and possesses certain substantial ecological significance (Bull and

L. K. Edison
Rajiv Gandhi Centre for Biotechnology (RGCB), Thiruvananthapuram, Kerala, India

T. G. Nandu
SpiceHealth COVID-19 RT-PCR Mobile Laboratory, Kasaragod, Kerala, India

N. S. Pradeep (✉)
KSCSTE-Malabar Botanical Garden and Institute for Plant Sciences, Kozhikode, Kerala, India

N. S. Pradeep, L. K. Edison (eds.), *Microbial Beta Glucanases*, Interdisciplinary Biotechnological Advances, https://doi.org/10.1007/978-981-19-6466-4_9

Chesters 1966; Mann et al. 1978). Considering the diverse applications of beta-glucanases, the current study is undertaken for the in silico analysis of enzymes at genomic and proteomic levels. It is enormously fascinating to distinguish the function of enzymes by interpreting the structural features. Protein motifs of beta-glucanases have been annotated in the Carbohydrate-Active Enzymes (CAZy) modules as sequence-related families (Blackman et al. 2014), and the activities have been annotated according to sequence similarities, protein structural folding, and recognized enzyme activities (Cantarel et al. 2009). A better understanding of the molecular structure of enzymes is essential as it helps to discover the relationship with functional properties. The molecular structure predominantly controls enzymes' physical and chemical properties, which retain their roles in nature and biological systems (Bielecki and Galas 1991).

Bioinformatics applications deliver more insights into sequence-based analysis and molecular level structure prediction of proteins. Computational biology paving the techniques to predict the three-dimensional structure and providing functional understandings will permit the required modifications applicable in many industrial applications. Several beta-glucanase sequences have been resolved, and their comparison has been used for elucidating the functional and structural relationships (Abraham et al. 2013; Lim et al. 2016). The knowledge from molecular docking studies allows us to predict the binding affinity and nature of the bond between two molecules, especially the candidates in drug design. In enzyme studies, it can be applied to predict the catalytic domains for binding substrate molecules (Lanka et al. 2015). Many three-dimensional structures of beta-glucanases of prokaryotic and eukaryotic origins are available in different databases (Hrmova and Fincher 2001); amongst them, structures of fungi and bacteria are more prominent. However, studies in *Streptomyces* spp. are rarely reported (George et al. 2010; Da Vinha et al. 2011).

9.2 Beta-Glucanases from Streptomyces

Hydrolysis of biological materials such as polysaccharides is highly challenging from an industrial viewpoint because of the unavailability of powerful enzymes (Kim et al. 2006). Hence, glycosyl hydrolase-secreting microorganisms have a significant impact on industrial sectors. It degrades complex sugars into simple ones (Pinheiro et al. 2017). However, a limited number of microorganisms produce efficient glycosyl hydrolyze (Lynd et al. 2002). It has been reported that *Streptomyces* spp. is a distinguished microorganism with efficient carbohydrate hydrolyzing ability. Hence, it is widely used in industrial applications (Book et al. 2016; Hoang et al. 2010). Soil inhabitant *Streptomyces* spp. are plant growth promoters and also provide disease-suppressive soils (Heinsch et al. 2019). Members within the *Streptomyces* genus have been widely exploited for searching for naturally occurring new compounds, predominantly antibiotics. Soil isolates are appreciated as principal producers of extracellular enzymes. The hydrolysis of carbohydrate substrates in

Streptomyces is controlled by innumerable carbohydrate-active enzymes (CAZymes); most are extracellular (Lombard et al. 2014).

According to Malik et al. 2020, the existence of strain-specific carbohydrate-active enzyme genes indicated the distinctiveness of the genetic makeup of *Streptomyces yeochonensis* strain CN732. Beta-glucanases are glycosyl hydrolases or carbohydrate-active enzymes (CAZyme). However, the present understanding of the glucano-lytic enzymes from *Streptomyces* is minimal. As per our prior studies, two soil isolates, *Streptomyces althioticus* TBG-MR17 and *Streptomyces cinereoruber* subsp. *cinereoruber* TBG-AL13, were illustrious as potent producers of extracellular exo-β-1,4-glucanase and endo-β-1,3-glucanase respectively. Therefore, the isolated complete gene sequences of exo-β-1,4-glucanase (EXO14) from *Streptomyces althioticus* TBG-MR17 and endo-β-1,3-glucanase (ENDO13) from *Streptomyces cinereoruber* subsp. *cinereoruber* TBG-AL13 were evaluated to predict the protein three-dimensional structure and their interaction with substrates based on an insilico approach (Edison et al. 2020a).

9.3 Beta-Glucanase Encoding Gene Isolation

The isolation and characterization of beta-glucanase encoding genes are the first phase for large-scale heterologous industrial production strategies. Identification of genes and in-depth analysis of gene sequence open new avenues to functional and comparative genomics, extending up to genomic and proteomic level engineering of enzymes. Understanding of sequence and structural aspects of beta-glucanase genes insight into the application of various gene expression technologies, site-directed mutagenesis, which modifies enzymatic effectiveness, cost-effective production, and enzyme stability in wide ranges of conditions (Handfield and Levesque 1999). Gene isolation strategy from individual strains includes DNA extraction, PCR amplification, cloning of isolated genes into suitable vectors, screening for positive clones on indicator medium, and sequencing of positive clones to identify the target gene (Lam et al. 2015). This is a very effective approach for isolating a beta-glucanase gene from pure bacterial strains without overexpression of protein as the overexpression is time consuming, and protein loss may affect the catalytic activity. The genome sequence information generated through high-throughput sequencing is deposited in databases that enable us to annotate the newly isolated beta-glucanase genes.

Evolutionary analysis of individual strains before gene isolation is recommended for retrieving the conserved region sequences for designing the primers. Gene sequences of phylogenetic neighbouring strains are similar. Then, the partially amplified genes are subjected to inverse PCR to obtain full-length sequences. The primers for the isolation of Exo-β-1,4-glucanase (EXO14) and endo-β-1,3-glucanase (ENDO13) genes from *Streptomyces althioticus* TBG-MR17 and *Streptomyces cinereoruber* sub sp. *cinereoruber* TBG-AL13 (the Western Ghats isolates) were designed based on the available gene sequences present in the phylogenetic neighbouring species of NCBI database. Those were aligned, and the primers were

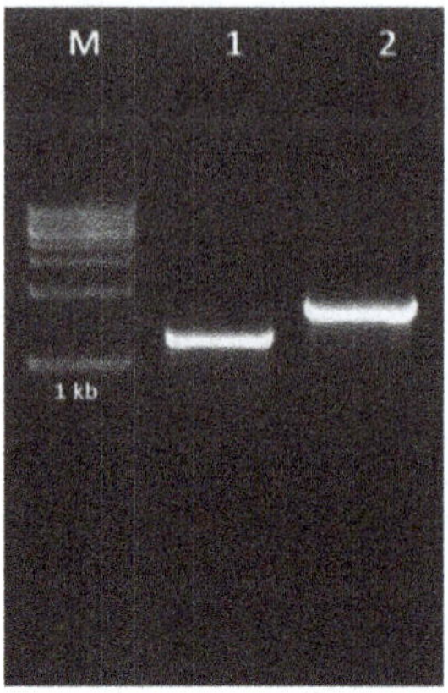

Fig. 9.1 PCR amplification of endo-β-1,3-glucanase and exo-β-1,4-glucanase genes. Lane M: 1 kb DNA ladder, Lane 1: endo-β-1,3-glucanase gene from *Streptomyces cinereoruber* subsp. *cinereoruber* TBG-AL13 yielded ~1400 bp and Lane 2: exo-β-1,4-glucanase gene from *Streptomyces althioticus* TBG-MR17 yielded ~2000 bps

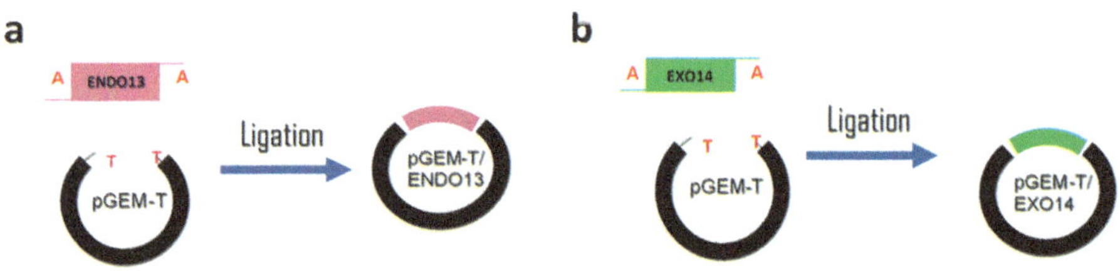

Fig. 9.2 Cloning of endo-β-1,3-glucanase and exo-β-1,4-glucanase genes in pGEM®-T Easy Vector. (**a**) Construction of pGEMT/ENDO13 plasmid. (**b**) Construction of pGEMT/EXO14 plasmid

picked from conserved regions of the aligned sequences. The PCR amplification provided a 1737 bp complete gene sequence of Exo-β-1,4-glucanase, which encoded a putative protein of 578 amino acids. The complete sequence of the endo-β-1,3-glucanase gene has 1293 bp, which is encoded by a putative protein of 438 amino acids (Fig. 9.1). The TA cloning vectors like pGEM®-T Easy Vector are used to cloning purified PCR products (Fig. 9.2). The cloned products were annotated after sequence analysis (Edison et al. 2020b).

9.4 Sequence Analysis of Beta-Glucanase Genes

The exploration of available genetic resources like genome and protein databases helps annotate the isolated enzyme encoding genes. Gene sequence contains a plethora of information such as protein sequence, Physicochemical properties (molecular weight, pI, stability, etc.), protein structural characteristics, substrate specificity, and conversion capability (Geoffrey et al. 2011). Comprehensive computational approaches and organized databases offer great potential to unravel the sequence–functional relationships of proteins. Protein sequences acquired from single gene sequences are available in public databases such as NCBI or UniProtKB. The functional annotation of the target enzyme coding gene was done by sequence comparison with already available annotated enzymes.

The open reading frames (ORF) of the target beta-glucanase gene sequence can be determined after sequence assembling and vector screening. CAP contig programme of BIOEDIT v7.2.5 helps to assemble the gene sequences (Hall 1999). VecScreen tool from NCBI (https://www.ncbi.nlm.nih.gov/tools/vecscreen/) allows separating the vector sequence from assembled gene sequences. The ExPASy Translate tool (http://web.expasy.org/translate/) translates the nucleotide sequences to protein sequences. BLASTp algorithm (http://blast.ncbi.nlm.nih.gov/Blast.cgi) with default settings from NCBI GenBank database aids to find the homologous sequences of the target protein sequence. Sequence alignments can be carried out with MAFFT v7 (http://align.bmr.kyushu-u.ac.jp/mafft/online/server/) with the E-INS-I algorithm (Katoh and Standley 2013). ESPript 3.0 (Robert and Gouet 2014) permits the aligned sequences view. The phylogenetic trees of a beta-glucanase encoding protein sequence can be constructed using the Neighbour-Joining algorithm in MEGA X (Kumar et al. 2018). Based on the multiple sequence alignment, BLASTp analysis, and phylogenetic tree construction confirmed Exo-β-1,4-glucanase protein from *Streptomyces althioticus* TBG-MR17 has a very close resemblance to the *Streptomyces* sp. 4F cellulose 1,4-β-cellobiosidase (Fig. 9.3a) and endo-β-1,3-glucanase protein *Streptomyces cinereoruber* sub sp. *cinereoruber* TBG-AL13 showed more similarity to *Streptomyces globisporus* endo-β-1,3-glucanase (Fig. 9.3b) (Edison et al. 2020b).

Significant features of the functional annotation of enzymes include both enzyme activity and substrate specificity, which can differ extensively within the family. The "sequence identity" strategy, such as pairwise protein sequence alignment, validates the functional diversity, catalytic site, and enzyme families. In most cases, amino acid sequence allows for predicting enzyme family and the substrate specificity, based on the adjacent neighbour classification rule. Combining structural evidence and sequence data can expand the interpretability and precision of the active site amino acids (Röttig et al. 2010). Understanding the role of carbohydrate-active enzymes is essential to recognize their industrial and biological role potential. The advances in sequence analysis put forward two hypotheses: conserved peptide patterns decide the protein similarities, and the second is spacing between adjacent amino acids. This approach enables the grouping of enzymes into families and/or subfamilies. Carbohydrate-active enzyme ANnotation (dbCAN) is a functional annotation tool of individual carbohydrate-active enzymes that uses conserved unique peptide patterns. Thus, the objective of developing the conserved (CUPP) method is to construct a peptide-based grouping and validate the method to provide evidence that CUPP captures function-related features of the (as defined by CAZy families). A standardized, efficient, and robust approach to functional annotation of carbohydrate-active enzymes would support improved molecular insight into enzyme–substrate interaction. It automatically interprets carbohydrate-active enzymes' protein, subfamily, and enzyme classification (Barrett and Lange 2019).

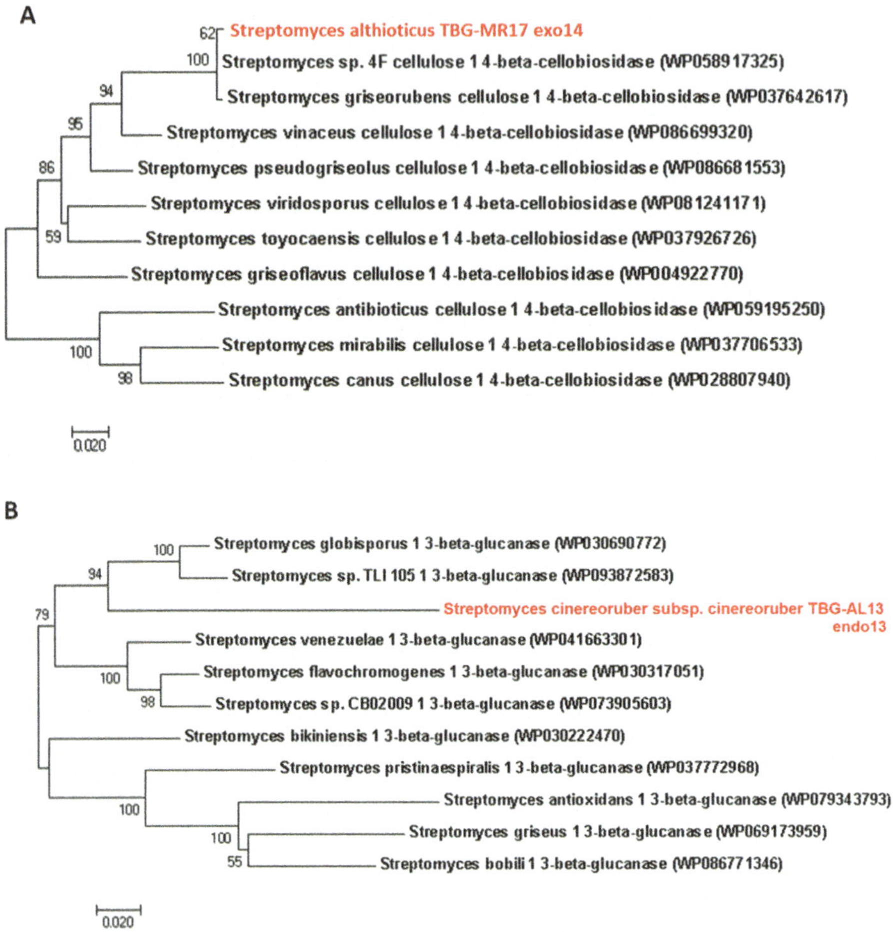

Fig. 9.3 Phylogenetic analysis of (**a**) *Streptomyces althioticus* TBG-MR17 exo-β-1,4-glucanase (exo14) and (**b**) *Streptomyces cinereoruber* subsp. *cinereoruber* TBG-AL13 endo-β-1,3-glucanase protein. The evolutionary history was inferred using the Neighbor-Joining method. The percentage of replicate trees in which the associated taxa clustered together in the bootstrap test (1000 replicates) are shown next to the branches. The tree is drawn to scale, with branch lengths in the same units as those of the evolutionary distances used to infer the phylogenetic tree. The evolutionary distances were computed using the Poisson correction method and are in the units of the number of amino acid substitutions per site. Evolutionary analyses were conducted in MEGAX

Conserved domain database (CDD) search is another annotation resource for full-length proteins and domains. It comprises a group of well-annotated multiple sequence alignment models.

dbCAN domain annotation (Yin et al. 2012) and CDD search (Marchler-Bauer et al. 2013) confirmed the exo-β-1,4-glucanase (exo14) from *Streptomyces althioticus* TBG-MR17 has N-terminal CBM2 domains and C-terminal GH6 domain (Edison et al. 2020b), the characteristics functional domains of enzyme

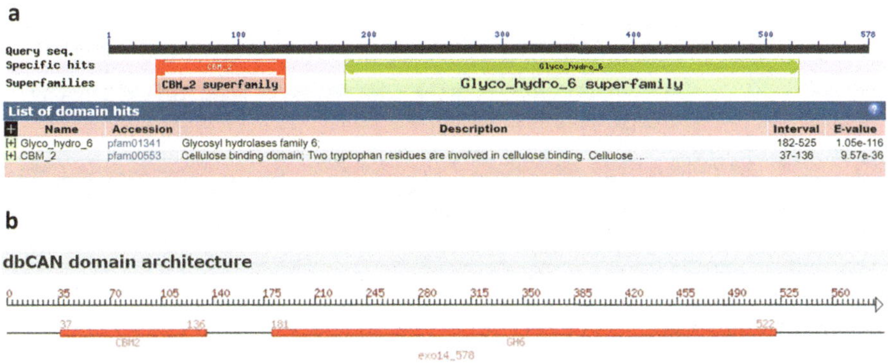

Fig. 9.4 Conserved domain search of exo-β-1,4-glucanase (exo14) from *Streptomyces althioticus* TBG-MR17. (**a**) NCBI Conserved domain search based on sequence alignment recognized exo14 has a linker which separates N-terminal CBM2 family and C-terminal GH6 family domains. (**b**) dbCAN domain search also confirmed the above domain architecture and the protein belongs to GH6 CAZyme family

exo-β-1,4-glucanase (cellulose 1,4-β-cellobiosidase or cellobiohydrolase) (Fig. 9.4) (Leo et al. 2018). According to Tomotsune et al. (2014), cellulose-producing *Streptomyces* have been allotted to the protein families of GH5, GH6, GH9, GH12, and GH48 in CAZy database. Annotation of the complete genome sequence of a pine-boring wood wasp associated *Streptomyces* sp. SirexAA-E (ActE) predicted 1,4-β-cellobiohydrolase with GH6 and CBM2 domains, but structural information is not elucidated (Takasuka et al. 2013). GH modules are also accompanied by CBM modules (Lopez-Casado et al. 2008; Häkkinen et al. 2012). Mostly CBM2 was fused on the C-terminal side of the catalytic domain in all glycoside hydrolase families, whereas in GH5 and GH6 domain families is located on the N-terminal end of domain architecture (Teegardin et al. 2017). Both GH and CBM domains are separated by a linker sequence, rich with Pro/Thr residues (Tomme et al. 1995; Jensen et al. 2018). In CBM2/GH6, the average length of the linker is 35 (Sammond et al. 2012). In our case, a linker with 45 amino acids is found in between these two domains. However, a major divergence from other sequences confirmed the endo-β-1,3-glucanase (endo13) *Streptomyces cinereoruber* sub sp. *cinereoruber* a novel one. Functional annotation of endo13 protein confirmed that it has an N-terminal Ricin-B-lectin-like GH16 and C-terminal CBM13 domains, typical for endo-β-1,3-glucanase enzymes. Both domains are separated by a 16 amino acid long linker sequence (Fig. 9.5) (Edison et al. 2020b). GH16 is one of the most prominent CAZyme families, including endoglucanases involved in glucan degradation (Temuujina et al. 2012). Endo-1,3-β-glucanase from *Cellulosimicrobium cellulans* reported an N-terminal GH16 catalytic domain and C-terminal CBM13 regions separated by a Gly/Ser-rich linker (Tanabe et al. 2008).

As per reports of Gilkes et al. (1991), the catalytic site of Exo-β-1,4-glucanase (cellulose 1,4-β-cellobiosidase) has a key characteristics consensus pattern, of V-x-

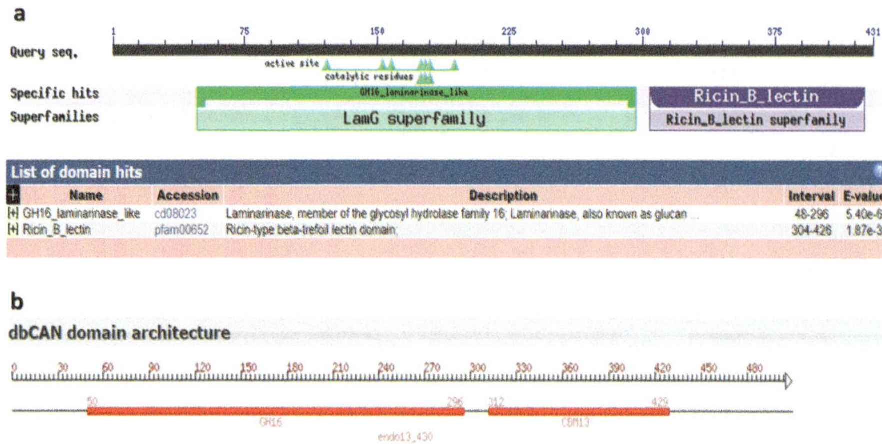

Fig. 9.5 Conserved domain search of endo-β-1,3-glucanase (endo13) from *Streptomyces cinereoruber* sub sp. *cinereoruber* TBG-AL13 (**a**) NCBI Conserved domain search based on sequence alignment recognized endo13 has N-terminal GH16-laminarinase-like domain and C-terminal Ricin-B-lectin family domains, both are separated by a linker sequence. (**b**) dbCAN Domain architecture of endo13 confirmed, the protein belongs to GH16 CAZyme family and Ricin-B-lectin domain was a carbohydrate-binding module 13 (CBM13) family of CAZy database

Y-x(2)-P-x-R-D-C-[GSAF]-x(2)-[GSA](2)-x-G. The exo14 protein alignment also showed a similar catalytic core pattern, VIYNLPGRDCAALASNG, between amino acids 236 and 254 (Fig. 9.6). Catalytic region of endo13 protein showed a consensus sequence WPNSGEIDVMENVGFEP, similar to the characteristic consensus pattern, W-P-x(2)-G-E-[LIV]-D-[LIVF]-x(0,1)-E-x(2)-[GQ]-[KRNF]-x-[PSTA], found in all other endo-β-1,3-glucanase (laminarinase) proteins (Fig. 9.7) (Edison et al. 2020b). The two glutamic acid residues, Glu-175 and Glu-180, within this conserved region are thought to be involved in catalysis (Juncosa et al. 1994), though the aspartic acid residue, Asp-177, is significant in preserving the relative position of the above catalytic amino acids (Michel et al. 2001). A minor difference was revealed inactive site sequences, WPKCGEIDIMELLG, of *Clostridium thermocellum* endo-β-1,3-glucanase, Lic16A (Fuchs et al. 2003). Our *Streptomyces* endo13 also showed some differences in the catalytic site regions of Laminarinase genes from *Rhodothermus marinus, Thermotoga neapolitana, Bacillus circulans*, and *Strongylocentrotus purpuratus* (Krah et al. 1998).

The secretion of hydrolytic enzymes into the culture supernatant has some significant benefits related to intracellular expression strategies. Extracellular secretion simplifies their downstream processing and remarkably reduces the production outlays. For extracellular secretion, protein has to be exported across the cytoplasmic membrane. In bacteria, two pathways are used for protein transport—the general secretion (Sec) pathway and the twin-arginine translocation (Tat) pathway. N-terminal signal peptides facilitate membrane translocation with the help of protein translocases. In bacteria, shortly after the cytoplasmic membrane transport, the

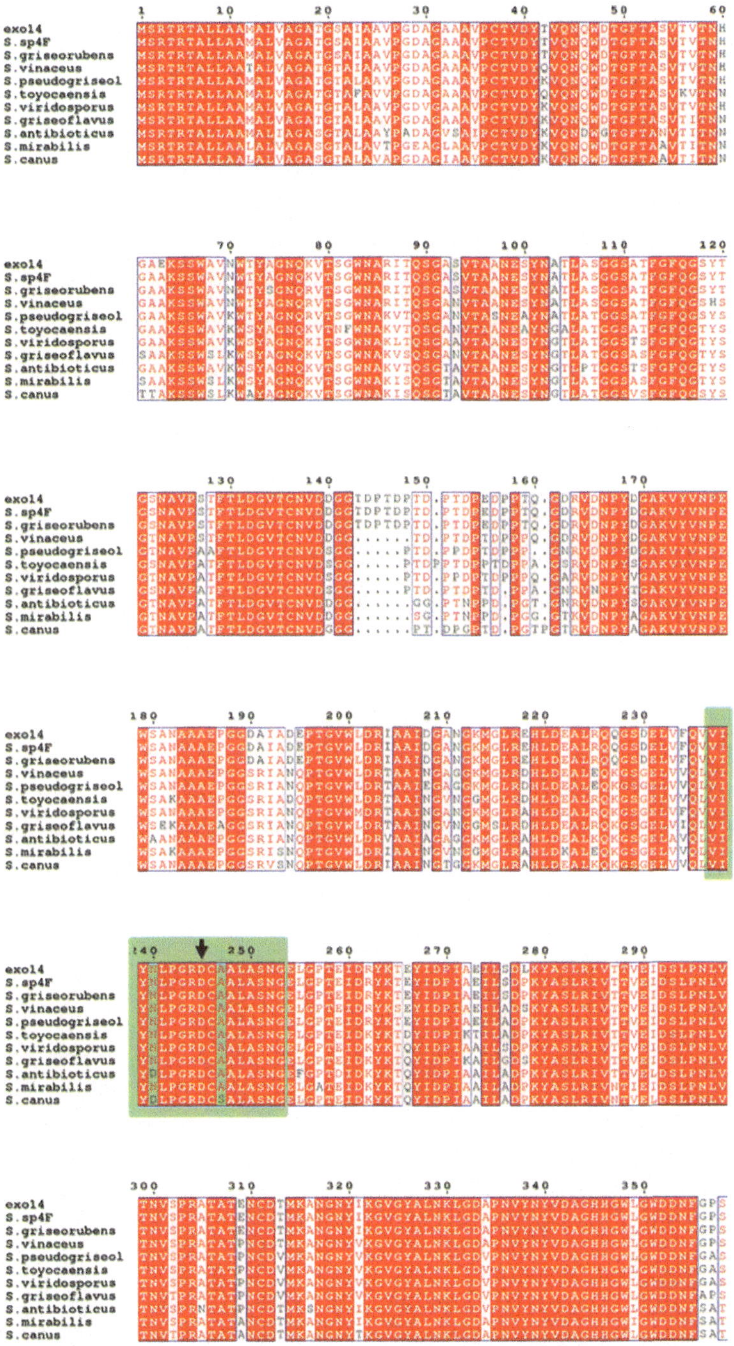

Fig. 9.6 Multiple sequence alignment of exo-β-1,4-glucanase (exol4) from *Streptomyces althioticus* TBG-MR17. Conserved residues are shown in dark shaded boxes and light shaded boxes indicated similar residues. Conserved active site regions of exo-β-1,4-glucanase protein are specified in green box. The catalytic amino acid residues are indicated by arrows

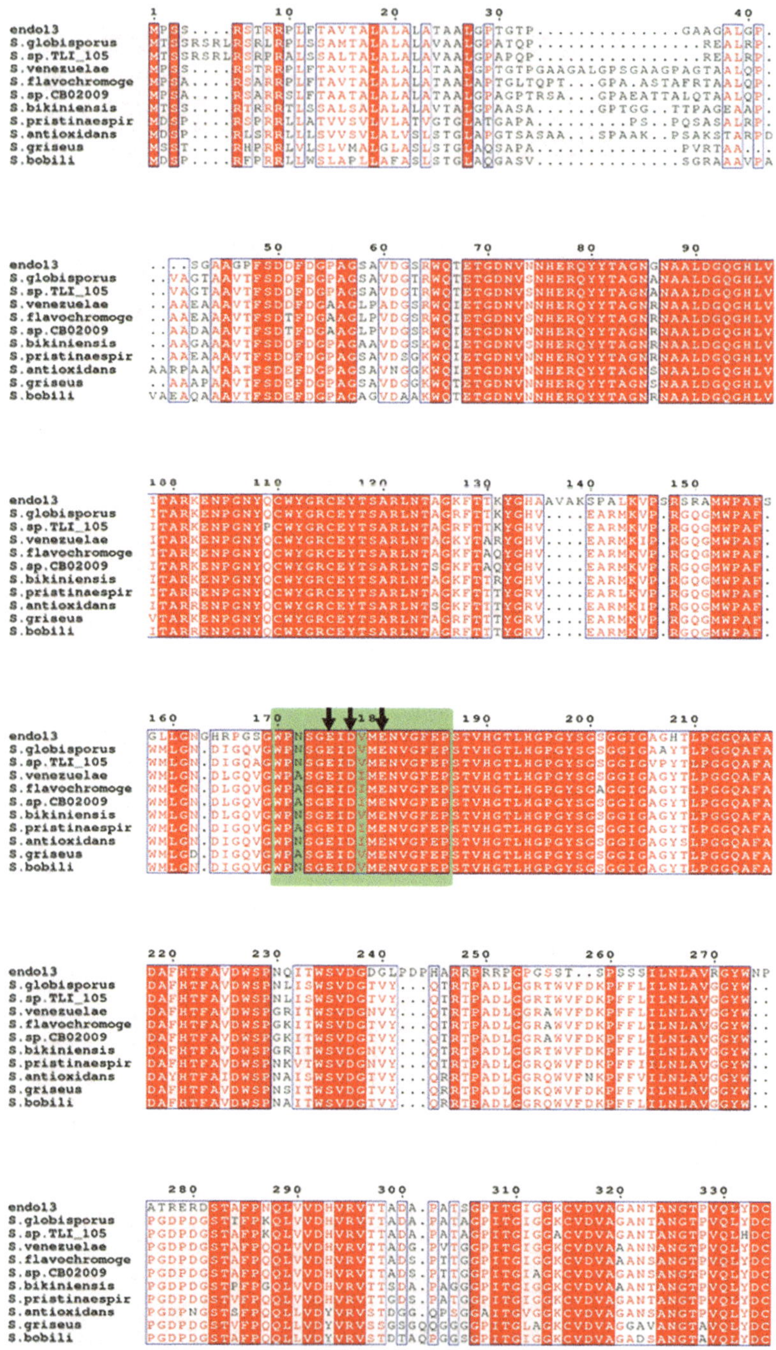

Fig. 9.7 Multiple sequence alignment of endo-β-1,3-glucanase (endo13) from *Streptomyces cinereoruber* sub sp. *Cinereoruber* TBG-AL13. Conserved residues are shown in dark shaded boxes and light shaded boxes indicated similar residues. Conserved active site regions of endo-β-1,3-glucanase protein are specified in green box. The catalytic amino acid residues are indicated by arrows

signal peptide is cleaved by enzyme signal peptidase, and the mature protein is released into the periplasmic space (Freudl 2018). Signal peptide and the signal peptidase cleavage sites can be predicted in silico by SignalP v4.1 (http://www.cbs.dtu.dk/services/SignalP/). SignalP programme detected the Exo-β-1,4-glucanase (exo14) from *Streptomyces althioticus* TBG-MR17 has a TAT signal peptide sequence of 32 amino acids in length, enhancing the efficiency of protein cleaves the mature protein in between Ala-32 and Ala-33. The TAT pathway mainly secretes proteins in a native folded state, found in *Streptomyces* and *E. coli* (Ize et al. 2003; Anne et al. 2014). Thirty-five amino acids long N-terminus signal peptide domain was found in endo-β-1,3-glucanase (endo13) *Streptomyces cinereoruber* sub sp. *cinereoruber*, cleaves in between Ala-35 and Ala-36 positions (Fig. 9.8) (Edison et al. 2020b).

9.5 In Silico Prediction of Physicochemical Parameters and Secondary Structures

EXPASY tool ProtParam (http://web.expasy.org/protparam/) computes several Physicochemical parameters like molecular weight, theoretical isoelectric point (pI), half-life, instability index, number of positive (+R) and negative residues (−R), aliphatic index, extinction coefficient, and grand average of hydropathicity (GRAVY) values (Gasteiger et al. 2005). Physicochemical property analysis showed the isoelectric point (pI) of exo-β-1,4-glucanase (exo14) from *Streptomyces althioticus* TBG-MR17 was 4.34, provided the possibility to develop a buffer system for pH-based purification (Abraham et al. 2013). At pI, proteins are compact and stable (Sahay and Shakya 2010). The estimated half-life of exo14 in *E. coli* was more than 10 h and the designated instability index value of 24.65 predicted the protein was stable. The high aliphatic value of 66.26 showed that the protein was stable at a varying range of temperatures since the aliphatic index is directly related to the thermostability of protein (Ikai 1980). Very low GRAVY value −0.446 concludes that the protein makes good interaction with water molecules. Physicochemical properties deduce that the endo-β-1,3-glucanase (endo13) *Streptomyces cinereoruber* sub sp. *cinereoruber* TBG-AL13 has a pI value of 8.1 (pI>7.0), revealed the basic nature of the protein. The estimated half-life in *E. coli* was more than 10 h and the protein was predicted as stable with an instability index of 29.37. The aliphatic value of 59.35 showed that the endo13 was less thermostable than exo14. A low GRAVY value, −0.424 revealed a good hydrophilic interaction (Table 9.1) (Edison et al. 2020b).

Protein secondary structure refers to the local conformation of the polypeptide backbone. Prediction of protein secondary structure delivers an essential step to tertiary structure prediction and provides information about activity function and

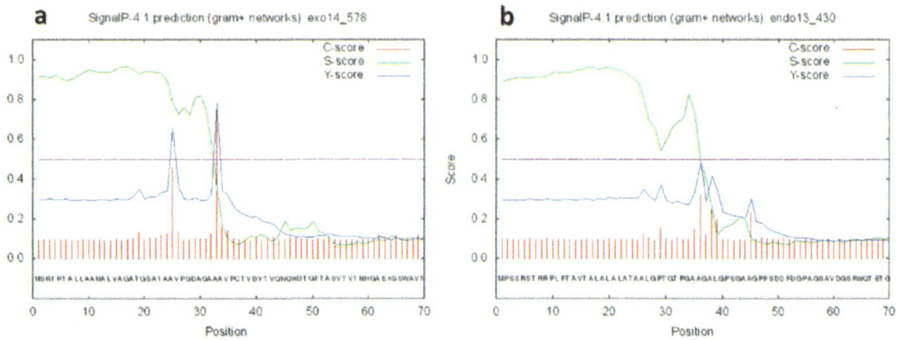

Fig. 9.8 Signal peptide prediction of exo-β-1,4-glucanase (exo14) from *Streptomyces althioticus* TBG-MR17 and endo-β-1,3-glucanase (endo13) from *Streptomyces cinereoruber* sub sp. *cinereoruber* TBG-AL13. (**a**) Predicted signal sequence of exo14 protein showed a cleavage site between A-32 and A-33 with a cutoff value of 0.450. (**b**) Predicted signal sequence of exo14 protein showed a cleavage site between A-35 and A-36 residues with a 0.450 cutoff

Table 9.1 Physicochemical properties of proteins exo-β-1,4-glucanase (exo14) from *Streptomyces althioticus* TBG-MR17 and endo-β-1,3-glucanase (endo13) from *Streptomyces cinereoruber* sub sp. *Cinereoruber* TBG-AL13

Parameters	exo14	endo13
No. of amino acid residues	578	430
Molecular weight (Daltons)	61152.67	44360.76
Chemical formula	$C_{2661}H_{4046}N_{742}O_{892}S_{14}$	$C_{1918}H_{2956}N_{582}O_{615}S_{11}$
Theoretical PI	4.34	8.11
+R residues	38	33
-R residues	76	31
Extinction coefficient	111,270–111,645	79,410–79,910
Instability index	24.65	29.37
Aliphatic index	66.26	59.35
Estimated half-life in *E. coli*	>10 hours	>10 hours
GRAVY	−0.446	−0.424

relationship. Each protein has three different secondary structure states-, and. α-helix (H) and β-strand (E) are the types of irregular secondary structures. The coil region (C) is an irregular secondary structure type. The secondary structure prediction of protein is generally assessed by Q3 accuracy, which measures the percentage of residues for three-state secondary structures to determine whether they have been predicted accurately (Ma et al. 2018). Protein topology or secondary structure predictions were performed with high confidence using the PSIPRED v3.3 (McGuffin et al. 2000). The secondary structure of exo-β-1,4-glucanase (exo14) from *Streptomyces althioticus* TBG-MR17 has 11 helixes and 15 sheets (Fig. 9.9a). Predicted secondary structure of endo-β-1,3-glucanase (endo13) *Streptomyces cinereoruber*

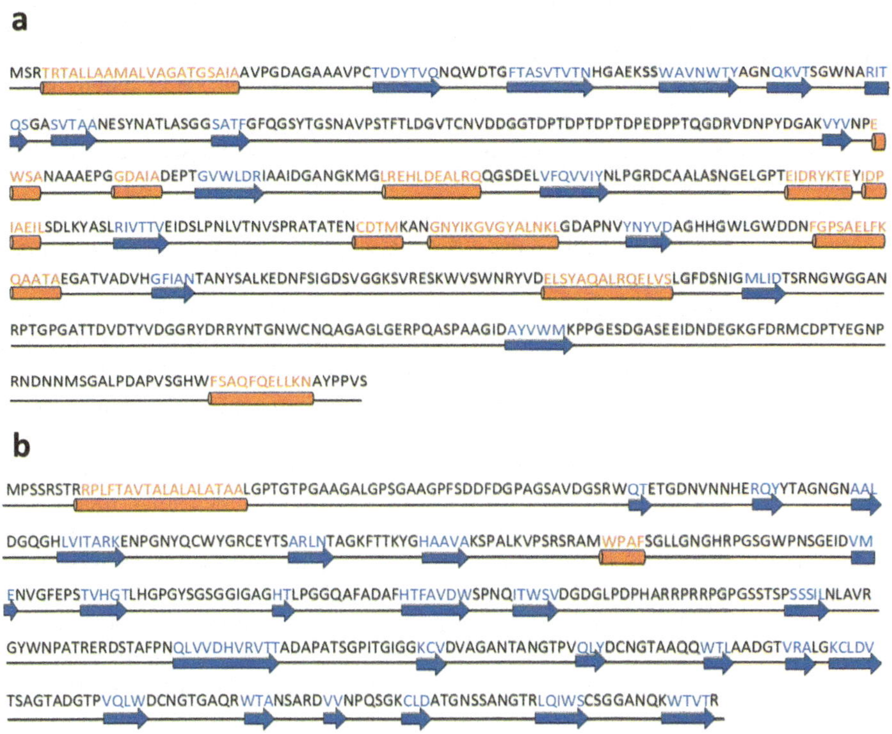

Fig. 9.9 Predicted secondary structure of exo-β-1,4-glucanase (exo14) from *Streptomyces althioticus* TBG-MR17 and endo-β-1,3-glucanase (endo13) from *Streptomyces cinereoruber* sub sp. *cinereoruber* TBG-AL13 by PSIPRED v3.3. (**a**) Predicted secondary structure of exo14 protein have 11 helixes and 15 sheets. (**b**) Predicted secondary structures ofendo13 have only 2 helixes and 24 sheets. α-helixes indicated in orange and β-sheets in blue

sub sp. *cinereoruber* TBG-AL13 has only 2 helixes and 24 sheets (Fig. 9.9b) (Edison et al. 2020b). Greater β-sheet content above other consistent secondary structures is a collective feature of β-1,3 glucanases (Abraham et al. 2013).

9.6 Three-Dimensional Structure Prediction

The complete analysis of the primary sequence leads to the prediction of new enzymes from already known families. Three-dimensional structure analysis of proteins can be predicted using a protein database called Protein Data Bank (RSCB PDB). It is a database catalogue of three-dimensional structural information of biological macromolecules, primarily proteins. The structure-based approach focusing on the highly conserved regions leads to annotating proteins towards predicting active site amino acids, substrate-binding efficiency, and catalytic properties. The primary step towards the three-dimensional structure prediction is finding active site topology and catalytic amino acid residues (Zaparucha et al. 2018; Zanghellini 2014).

Three-dimensional models of Exo-β-1,4-glucanase (exo14) from *Streptomyces althioticus* TBG-MR17 and endo-β-1,3-glucanase (endo13) *Streptomyces cinereoruber* sub sp. *cinereoruber* TBG-AL13 were generated using I-TASSER web server of the University of Michigan, the USA maintained by Zhang Lab, which uses the treading technique (Roy et al. 2010). I-TASSER used multiple threading approaches for selecting the templates and predicted full-length models of proteins by iterative structure assembly. Based on Z-score, one template selects from each threading programme to predict the target model of each protein. It generates 5 models by clustering the stimulated structural confirmations based on pairwise structural similarity with the template. The three-dimensional models are then evaluated based on TM-score (template modelling Score), C-score (confidence score), and RMSD (Root-mean-square deviation). The confidence of models is measured quantitatively by C-score, generally in the range of −5 to 2. The higher C-score value signifies a high confidence model. The RMSD and TM-score are calculated mainly based on the C-score. Finally, using the TM-align structural alignment programme I-TASSER aligns the best model to all available PDB structures. Based on the closest structural similarity, the programme selects the 10 best structural analogues from the Protein Data Bank.

The predicted three-dimensional structures of exo14 and endo13 proteins using I-TASSER were proved as good quality with proper fold confirmation. Structural modelling in I-TASSER is based on the secondary structure enhanced Profile-Profile threading alignment (Deshmukh et al. 2016). Initially, it predicts secondary structure and then goes to tertiary structure. For exo14 protein modelling, 4b4fA (*Thermobifida fusca Cel6B*, Sandgren et al. 2013) was selected as the best template with Z-score 2.46, and model 1 was the best template selected as a top-ranked model based on a high C-score (−2.07), signified the model was good. This was additionally validated by the highest TM-Score (0.47 ± 0.15) and RMSD (12.7 ± 4.3 Å) (Fig. 9.10a). I-TASSER also identified the enzyme commission classification of exo14 protein, EC 3.2.1.91, using structure-based function annotation algorithm of COFACTOR. The molecular modelling of endo13 protein selected 3dgtA with Z-score = 8.44 (endo-1,3-β-glucanase from *Streptomyces sioyaensis*, Hong et al. 2008) was chosen as the best template, and the best model was built with a high C-score value 1.01, estimated TM-Score = 0.59 ± 0.14 and estimated RMSD = 9.3 ± 4.6 Å (Fig. 9.10b). The identified EC classification of endo13 by I-TASSER was EC 3.2.1.39. Further validation of predicted structures was done by the superimposition of their corresponding templates (Fig. 9.10c and d). The validation based on Ramachandran Plot also showed none of the individual amino acids except proline and glycine was in a bad packaging region, indicating the predicted model was good (Fig. 9.11a and b) (Edison et al. 2020b).

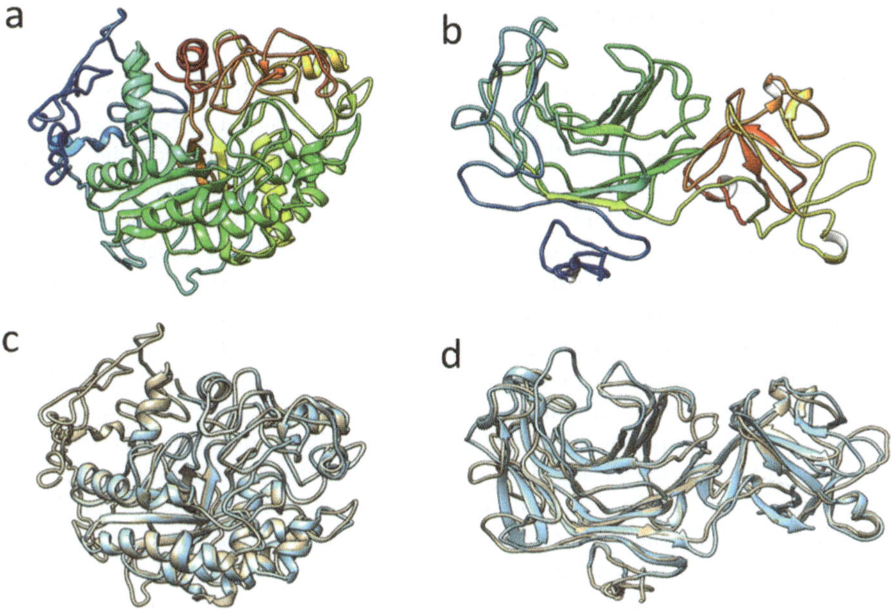

Fig. 9.10 Three-dimensional modelling of exo-β-1,4-glucanase (exo14) from *Streptomyces althioticus* TBG-MR17 and endo-β-1,3-glucanase (endo13) from *Streptomyces cinereoruber* sub sp. *cinereoruber* TBG-AL13. (**a**) Predicted 3D model of exo14 protein showed a tunnel-like catalytic pocket. (**b**) 3D model of predicted endo13 protein showed an open cleft-like catalytic pocket. (**c**) Superimposition of exo14 (beige colour) and 4b4fA (blue colour). (d) Superimposition of endo13 (beige colour) and 3dgtA (blue colour)

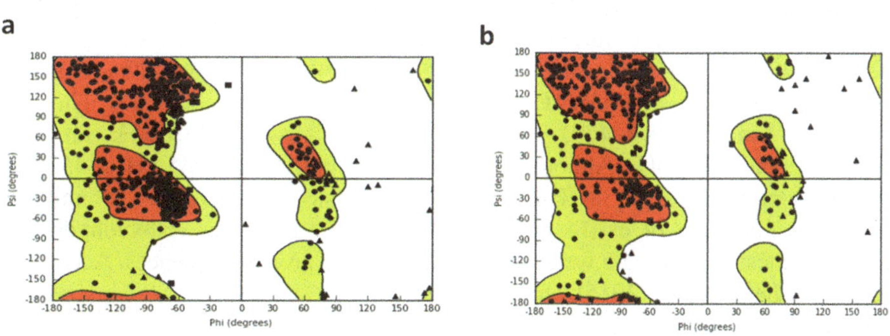

Fig. 9.11 Validation of 3-D models of exo-β-1,4-glucanase (exo14) from *Streptomyces althioticus* TBG-MR17 and endo-β-1,3-glucanase (endo13) from *Streptomyces cinereoruber* sub sp. *cinereoruber* TBG-AL13 using Ramachandran plot. (**a**) Ramachandran plot of exo14 protein. (**b**) Ramachandran plot of endo13 protein. Glycine is plotted as triangles, proline as squares, and other residues as circles. The orange regions are represented as "favoured" regions, the yellow regions are the "allowed" regions, and the white regions are the "disallowed" regions

9.7 Molecular Docking

Molecular docking is the commonly used method for studying the substrate specificity of an enzyme. The molecular docking of enzyme–substrate has three stages: protein (enzyme) preparation, ligand (substrate) preparation, and docking. In protein, preparation includes structural minimizations and optimizations. It also makes side chains and loops with missing atoms, adds hydrogens, optimizes the networks of H-bonding, and lastly, restrains minimization from getting a refined or superior protein structure for molecular interaction studies. The binding site within the catalytic domain is considered a substrate-binding site. Two-dimensional structures of ligand molecules in sdf format can be recovered from the PubChem compound database (https://pubchem.ncbi.nlm.nih.gov/) of NCBI. Ligand preparation includes adding hydrogen atoms, eliminating the inconsistencies between bond angle and length, and eventually converting the ligand to a three-dimensional structure. A receptor grid is created during docking in the expected active site's centre; ultimately, the prepared grid docks with the ligand's single finest conformation. Later, the finest pose of ligand to protein with determined docking score and docking energy is generated (Friesner et al. 2006).

Docking studies of exo-β-1,4-glucanase (exo14) from *Streptomyces althioticus* TBG-MR17 and endo-β-1,3-glucanase (endo13) from *Streptomyces cinereoruber* sub sp. *cinereoruber* TBG-AL13 were done using molecular docking software (Glide, Schrödinger Suites 2018–1). The exo14 used cellotriose (Fig. 9.12a) as a ligand. It showed a favourable interaction with active site furrow (site score 1.008, size 102, Dscore 1.052, volume 407.484) of protein and formed a better docking score of −7.393 and docking energy −79.575 (Fig. 9.12b). As Wu et al. (2013) mentioned. The study exposed the fact that the exo14 active site is a tunnel type. The significant amino acid residues involved in the binding of cellotriose were found to be Tyr-239, Asp-245, Ala-247, Ala-248, Ser-251, Asp-292, Asn-296, Thr-299, Asn-300, Thr-306, His-345, Gly-347, Trp-348, Trp-351, Asn-354, Thr-383, Ala-384, Asn-385, Asn-479, Thr-480, Gly-481, Trp-483, Met-533, Pro-542, Arg-543, and Asn-544. Although only three residues (Asp-245, His-345, and Arg-543) were involved in H-bonding interactions (Fig. 9.12c). The multiple sequence alignment of exo14 protein with other *Streptomyces* cellobiohydrolase found all these residues are well-maintained in all the protein sequences. According to Chinnathambi et al. (2015), cbhII putative catalytic site residues are conserved in different *Trichoderma* spp.

The molecular interaction studies of endo13 protein were performed with ligand laminaritriose (Fig. 9.13a). The active site consisted of a site score of 0.952, size 994, Dscore 1.070, and volume 749.455 was used for the molecular docking studies of endo13 protein. Here the active site is an open cleft-like structure, with more propensity to cut internal bonds (Wu et al. 2013). The docking results produced a

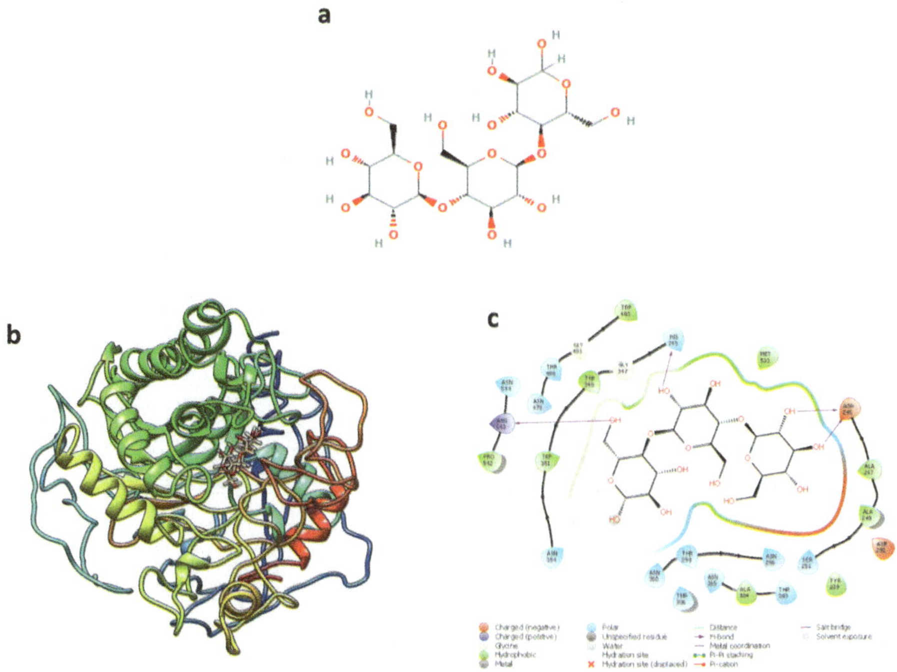

Fig. 9.12 (**a**) Chemical structure of substrate Cellotriose (PubChem CID: 5287993), Molecular Formula: $C_{18}H_{32}O_{16}$ and Molecular Weight: 504.438 g/mol. (**b**) Docked structures of exo14 and cellotriose molecule (ball and stick representation). (**c**) Ligand interaction diagram of exo14 protein showing the interaction of active site amino acid residues with cellotriose molecule

positive interaction at a better energy level with a docking score of −11.026 and model energy −78.365 (Fig. 9.13b). The detected amino acid residues involved in catalysis binding were Thr-69, Tyr-108, Tyr-112, Tyr-117, Thr-118, Ser-119, Ala-155, Ser-157, Leu-159, Leu-160, Asn-162, Ser-168, Trp-170, Ser-173, Gly-174, Glu-175, Asn-177, Pro-196, Ile-264, Asn-266, and Tyr-272. Six amino acid residues viz., Thr-118, Ser-157, Glu-175, Ser-173, Asn-162, and Trp-170 were actively involved in eight hydrogen bonding interactions (Fig. 9.13c).

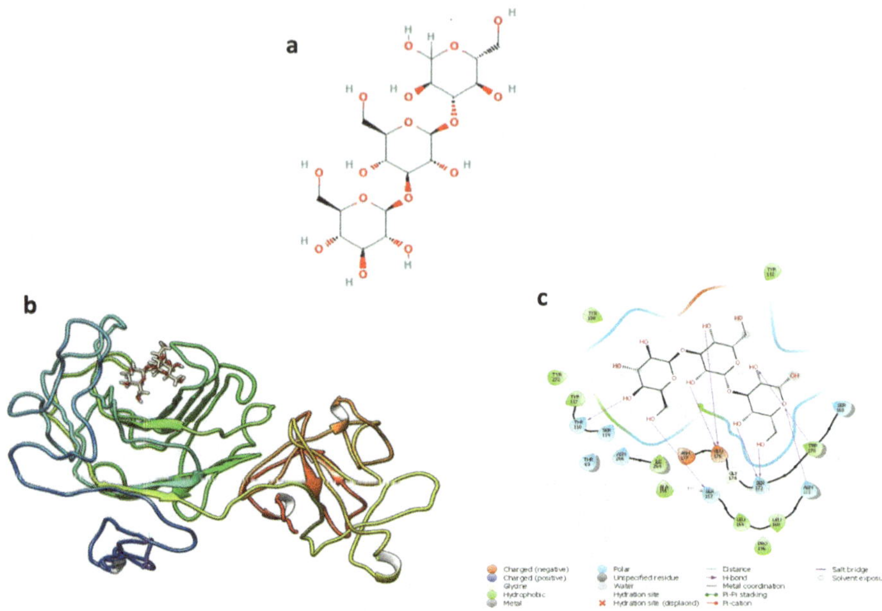

Fig. 9.13 (**a**) Chemical structure of Laminaritriose (PubChem CID:11477625), Molecular Formula: $C_{18}H_{32}O_{16}$ and Molecular Weight: 504.438 g/mol. (**b**) Docked structures of endo13 and laminaritriose molecule (ball and stick representation). (**c**) Ligand interaction diagram of endo13 protein showing the interaction of active site amino acid residues with laminaritriose molecule

References

Abraham A, Narayanan SP, Philip S, Nair DG, Chandrasekharan A, Kochupurackal J (2013) In silico characterization of a novel β-1,3-glucanase gene from *bacillus amyloliquefaciens*-a bacterial endophyte of *Hevea brasiliensis* antagonistic to *Phytophthora meadii*. J Mol Model 19(3):999–1007

Anne J, Vrancken K, Mellaert LV, Impe JV, Bernaerts K (2014) Protein secretion biotechnology in gram-positive bacteria with special emphasis on *Streptomyces lividans*. Biochim Biophys Acta 1843:1750–1761

Barrett K, Lange L (2019) Peptide-based functional annotation of carbohydrate-active enzymes by conserved unique peptide patterns (CUPP). Biotechnol Biofuels 12:102

Bielecki S, Galas E (1991) Microbial beta-glucanases different from cellulases. Crit Rev Biotechnol 10(4):275–304

Blackman LM, Cullerne DP, Hardham AR (2014) Bioinformatic characterisation of genes encoding cell wall degrading enzymes in the *Phytophthora parasitica* genome. BMC Genomics 15:785

Book AJ, Lewin GR, McDonald BR, Takasuka TE, Wendt-Pienkowski E, Doering DT, Suh S, Raffa KF, Fox BG et al (2016) Evolution of high cellulolytic activity in symbiotic *Streptomyces* through selection of expanded gene content and coordinated gene expression. PLoS Biol 14(6): e1002475

Bull AT, Chesters CGC (1966) The biochemistry of laminarin and the nature of laminarinase. Adv Enzymol Relat Areas Mol Biol 28:325–364

Cantarel BL, Coutinho PM, Rancurel C, Bernard T, Lombard V, Henrissat B (2009) The carbohydrate-active EnZymes database (CAZy): an expert resource for Glycogenomics. En. Nucleic Acids Res 37:D233–D238

Chinnathambi V, Balasubramanium M, Gurusamy R, Paramasamy G (2015) Molecular cloning and expression of a family 6 cellobiohydrolase gene cbhII from Penicillium funiculosum nCL-1. Adv Biosci Biotechnol 6:213–222

Da Vinha FN, Gravina-Oliveira MP, Franco MN, Macrae A, da Silva Bon EP, Nascimento RP, Coelho RR (2011) Cellulase production by Streptomyces viridobrunneus SCPE-09 using lignocellulosic biomass as inducer substrate. Appl Biochem Biotechnol 164(3):256–267

Deshmukh RA, Jagtap S, Mandal MK, Mandal SK (2016) Purification, biochemical characterization and structural modelling of alkali-stable β-1,4-xylan xylanohydrolase from aspergillus fumigatus R1 isolated from soil. BMC Biotechnol 16:11

Edison LK, Dan VM, Reji SR, Pradeep NS (2020a) A strategic production improvement of Streptomyces Beta glucanase enzymes with aid of codon optimization and heterologous expression. Biosci Biotechnol Res Asia 17(3):587–599

Edison LK, Reji SR, Pradeep NS (2020b) In silico perceptions in structural elucidation of Exo-β-1,4-glucanase (exo14) and endo-β-1,3-glucanase (endo13) from Streptomyces spp. J Appl Biol Biotechnol 8(6):67–81

Freudl R (2018) Signal peptides for recombinant protein secretion in bacterial expression systems. Microb Cell Factories 17(1):52

Friesner RA, Murphy RB, Repasky MP, Frye LL, Greenwood JR, Halgren TA, Sanschagrin PC, Mainz DT (2006) Extra precision glide: docking and scoring incorporating a model of hydrophobic enclosure for protein-ligand complexes. J Med Chem 49(21):6177–6196

Fuchs KP, Zverlov VV, Velikodvorskaya GA, Lottspeich F, Schwarz WH (2003) Lic16A of clostridium thermocellum, a non-cellulosomal, highly complex endo-b-1,3-glucanase bound to the outer cell surface. Microbiology (Reading) 149(4):1021–1031

Gasteiger E, Hoogland C, Gattiker A (2005) Protein identification and analysis tools on the ExPASy Server. In: The Proteomics Protocols Handbook, Humana Press, (ed) John M. Walker, pp 571–607

Geoffrey BA, Anke H, Santosh KP, Sebastian S, Uwe TB (2011) Discovery and protein engineering of biocatalysts for organic synthesis. Adv Synth Catal 353(13):2191–2215

George J, Arunachalam R, Paulkumar K, Wesely EG, Shiburaj S, Annadurai G (2010) Characterization and phylogenetic analysis of cellulase producing Streptomyces noboritoensis SPKC1. Interdiscip Sci 2(2):205–212

Gilkes NR, Claeyssens M, Aebersold R, Henrissat B, Meinke A, Morrison HD, Kilburn DG, Warren RAJ, Miller RC (1991) Structural and functional relationships in two families of beta-1,4-glycanases. Eur J Biochem 202(2):367–377

Häkkinen M, Arvas M, Oja M, Aro N, Penttilä M, Saloheimo M, Pakula TM (2012) Reannotation of the CAZy genes of Trichoderma reesei and transcription in the presence of lignocellulosic substrates. Microb Cell Factories 11:134

Hall TA (1999) BioEdit: a user-friendly biological sequence alignment editor and analysis program for windows 95/98/NT. Nucleic Acids Symp S 41:95–98

Handfield M, Levesque RC (1999) Strategies for isolation of in vivo expressed genes from bacteria. FEMS Microbiol Rev 23(1, January):69–91

Heinsch SC, Hsu SY, Otto-Hanson L, Kinkel L, Smanski MJ (2019) Complete genome sequences of Streptomyces spp. isolated from disease-suppressive soils. BMC Genomics 20(1):994

Hoang KC, Lai TH, Lin CS, Chen YT, Liau CY (2010) The chitinolytic activities of Streptomyces sp. TH-11. Int J Mol Sci 12(1):56–65

Hong TY, Hsiao YY, Meng M, Li TT (2008) The 1.5 a structure of endo-1,3-beta-glucanase from Streptomyces sioyaensis: evolution of the active-site structure for 1,3-beta-glucan-binding specificity and hydrolysis. Acta Crystallogr D Biol Crystallogr 64(9):964–970

Hrmova M, Fincher GB (2001) Structure-function relationships of β-D-glucan endo- and exohydrolases from higher plants. Plant Mol Biol 47(1–2):73–91

Ikai A (1980) Thermostability and aliphatic index of globular proteins. J Biochem 88(6):1895–1898
Ize B, Stanley NR, Buchanan G, Palmer T (2003) Role of the Escherichia coli tat pathway in outer membrane integrity. Mol Microbiol 48(5):1183–1193
Jensen MS, Fredriksen L, MacKenzie AK, Pope PB, Leiros I, Chylenski P, Williamson AK, Christopeit T, Østby H et al (2018) Discovery and characterization of a thermostable two-domain GH6 endoglucanase from a compost metagenome. PLoS One 13(5):e0197862
Juncosa M, Pons J, Dot T, Querol E, Planas A (1994) Identification of active site carboxylic residues in bacillus licheniformis 1,3-1,4-beta-D-glucan 4-glucanohydrolase by site-directed mutagenesis. J Biol Chem 269(20):14530–14535
Katoh K, Standley DM (2013) MAFFT multiple sequence alignment software version 7: improvements in performance and usability. Mol Biol Evol 30(4):772–780
Kim J, Yun S, Ounaies Z (2006) Discovery of cellulose as a smart material. Macromolecules 39(12):4202–4206
Krah M, Misselwitz R, Politz O, Thomsen KK, Welfle H, Borriss R (1998) The laminarinase from thermophilic eubacterium Rhodothermus marinus: conformation, stability, and identification of active site carboxylic residues by site-directed mutagenesis. Eur J Biochem 257(1):101–111
Kumar S, Stecher G, Li M, Knyaz C, Tamura K (2018) MEGA X: molecular evolutionary genetics analysis across computing platforms. Mol Biol Evol 35(6):1547–1549
Lam KN, Cheng J, Engel K, Neufeld JD, Charles TC (2015) Current and future resources for functional metagenomics. Front Microbiol 6:1196
Lamp AE, Evans AM, Moritz JS (2015) The effects of pelleting and glucanase supplementation in hulled barley based diets on feed manufacture, broiler performance, and digesta viscosity. J Appl Poult Res 24(3):295–303
Lanka S, Talluri VR, Ganesh J, Latha NL (2015) Homology modelling, molecular dynamic simulations and docking studies of a new cold active extracellular lipase. En:L A from Emericella nidulans NFCCI 3643. Trends Bioinformatics 8(2):37–51
Leo VV, Asem D, Zothanpuia, Singh BP (2018) Actinobacteria: A Highly Potent Source for holocellulose Degrading Enzymes. In: Singh BP et al (eds) Actinobacteria. New and future developments in microbial biotechnology and bioengineering. Elsevier, Amsterdam, pp 191–205
Lim JH, Lee CR, Dhakshnamoorthy V, Park JS, Hong SK (2016) Molecular characterization of Streptomyces coelicolor a (3). FEMS Microbiol Lett 363(3):245
Lombard V, Golaconda Ramulu H, Drula E, Coutinho PM, Henrissat B (2014) The carbohydrate-active enzymes database (CAZy) in 2013. Nucleic Acids Res 42(Database issue):D490–D495
Lopez-Casado G, Urbanowicz BR, Damasceno CM, Rose JK (2008) Plant glycosyl hydrolases and biofuels: a natural marriage. Curr Opin Plant Biol 11(3):329–337
Lynd LR, Weimer PJ, van Zyl WH, Pretorius IS (2002) Microbial cellulose utilization: fundamentals and biotechnology. Microbiol Mol Biol Rev 66(506):577. Erratum in: Microbiol Mol Biol Rev. (2002) 2066:2739
Ma Y, Liu Y, Cheng J (2018) Protein secondary structure prediction based on data partition and semi-random subspace method. Sci Rep 8(1):9856
Malik A, Kim YR, Jang IH, Hwang S, Oh DC, Kim SB (2020) Genome-based analysis for the bioactive potential of Streptomyces yeochonensis CN732, an acidophilic filamentous soil actinobacterium. BMC Genomics 21(1):118
Mann JW, Jeffries TW, Macmillan JD (1978) Production and ecological significance of yeast cell wall degrading enzymes from Oerskovia. Appl Environ Microbiol 36(4):594–605
Marchler-Bauer A, Zheng C, Chitsaz F, Derbyshire MK, Geer LY, Geer RC, Gonzales NR, Gwadz M, Hurwitz DI et al (2013) CDD: conserved domains and protein three-dimensional structure. Nucleic Acids Res 41(Database issue):D348–D352
McGuffin LJ, Bryson K, Jones DT (2000). The PSIPRED protein structure prediction server. Bioinformatics, 16, 404–405

Michel G, Chantalat L, Duee E, Barbeyron T, Henrissat B, Kloareg B, Dideberg O (2001) The kappa-carrageenase of P. carrageenovora features a tunnel-shaped active site: a novel insight in the evolution of clan-B glycoside hydrolases. Structure 9(6):513–525

Pinheiro GL, de Azevedo-Martins AC, Albano RM, de Souza W, Frases S (2017) Comprehensive analysis of the cellulolytic system reveals its potential for deconstruction of lignocellulosic biomass in a novel Streptomyces sp. Appl Microbiol Biotechnol 101(1):301–319

Robert X, Gouet P (2014) Deciphering key features in protein structures with the new ENDscript server. Nucleic Acids Res 42(Web Server issue):W320–W324

Röttig M, Rausch C, Kohlbacher O (2010) Combining structure and sequence information allows automated prediction of substrate specificities within enzyme families. PLoS Comput Biol 6(1): e1000636

Roy A, Kucukural A, Zhang Y (2010) I-TASSER: a unified platform for automated protein structure and function prediction. Nat Protoc 5(4):725–738

Sahay A, Shakya M (2010) In silico analysis and homology modelling of antioxidant proteins of spinach. J Proteomics Bioinform 03(5):148–154

Sammond DW, Payne CM, Brunecky R, Himmel ME, Crowley MF, Gregg GT, Beckham GT (2012) Cellulase linkers are optimized based on domain type and function: insights from sequence analysis, biophysical measurements, and molecular simulation. PLoS One 7(11): e48615

Sandgren M, Wu M, Karkehabadi S, Mitchinson C, Kelemen BR, Larenas EA, Ståhlberg J, Hansson H (2013) The structure of a bacterial cellobiohydrolase: the catalytic core of the Thermobifida fusca family GH6 cellobiohydrolase Cel6B. J Mol Biol 425(3):622–635

Takasuka TE, Book AJ, Lewin GR, Currie CR, Fox BG (2013) Aerobic deconstruction of cellulosic biomass by an insect associated Streptomyces. Science 3(1030):1–10

Tanabe Y, Pang Z, Oda M (2008) Cloning and sequencing of endo-1,3-β-glucanase from Cellulosimicrobium cellulans. J Biol Macromol 8(3):60–63

Teegardin KA, James S, Barabote RD (2017) Bioinformatic analysis of glycoside hydrolases in the proteomes of mesophilic and thermophilic Actinobacteria. MOJ Proteom Bioinformatics 5(3): 75–81

Temuujina U, Chia W, Chang YB, Honga S (2012) Identification and biochemical characterization of SCO3487 from Streptomyces coelicolorA3 (2), an exo- and endo-type–agarase producing neoagarobiose. J Bacteriol 194(1):142–149

Tomme P, Diane P, Emily DA, Robert AC, Miller JR, Antony R, Warren J, Kilburn DG (1995) Comparison of a fungal (family I) and bacterial (family II) cellulose-binding domain. J Bacteriol 177(15):4356–4363

Tomotsune K, Kasuga K, Tsuchida M, Shimura Y, Kobayashi M, Agematsu H, Ikeda H, Ishikawa J, Kojima I (2014) Cloning and sequence analysis of the cellulase genes isolated from two cellulolytic Streptomycetes and their heterologous expression in Streptomyces lividans. Int J Soc Mater Eng Resour 20(2):213–218

von Wettstein DV, Mikhaylenko G, Froseth JA, Kannangara CG (2000) Improved barley broiler feed with transgenic malt containing heat-stable (1,3-1,4)-β-glucanase. Proc Natl Acad Sci U S A 97(25):13512–13517

Wu M, Nerinckx W, Piens K, Ishida T, Hansson H, Sandgren M, Ståhlberg J (2013) Rational design, synthesis, evaluation and enzyme–substrate structures of improved fluorogenic substrates for family 6 glycoside hydrolases. FEBS J 280(1):184–198

Yin Y, Mao X, Yang J, Chen X, Mao F, Ying X (2012) dbCAN: a web resource for automated carbohydrate-active enzyme annotation. Nucleic Acids Res 40:445–451

Zanghellini A (2014) De novo computational enzyme design. Curr Opin Biotechnol 29:132–138

Zaparucha A, de Berardinis V, Vaxelaire-Vergne C (2018) Genome mining for enzyme discovery. In: Modern biocatalysis: advances towards synthetic biological systems. pp. 1-27

Chapter 10
Application of Codon Optimization and Recombinant Expression of Beta-Glucanases

Lekshmi K. Edison, Vipin Mohan Dan, and N. S. Pradeep

Abstract Beta-glucanases are industrially important glucan hydrolyzing enzymes. Beta-glucanase produced from *Streptomyces* sp. is more efficient and highly stable under varying temperature and pH conditions. However, large-scale production requires a huge quantity of biomass for sufficient enzyme production. An alternative approach for industrial production is the use of a heterologous expression system. Though, codon bias is the major difficulty in heterologous production, which significantly reduces the production. The effectiveness of the codon optimization strategy using synonymous substitution of codons increases protein expression levels during large-scale production. Codon optimization elevates the production of beta-glucanase by *Streptomyces* spp. genes in heterologous hosts and accomplished a profitable result on protein yield and functional activity. As a result of a multiparameter algorithm in gene optimization, the optimized and synthesized genes perform far better than the wild-type genes. The optimized construct directly influences mRNA stability and native folding of proteins, which increases enzyme activities. Altogether, gene optimization leads to hiked gene expression at various molecular processes like transcription and translation, resulting in higher product yield and improved activity.

Keywords Heterologous production · *Streptomyces* spp. · Codon adaptation index · Codon usage

L. K. Edison
Rajiv Gandhi Centre for Biotechnology (RGCB), Thiruvananthapuram, Kerala, India

V. M. Dan
School of Biochemistry and Immunology, Trinity College Dublin, Dublin, Ireland

N. S. Pradeep (✉)
KSCSTE-Malabar Botanical Garden and Institute for Plant Sciences, Kozhikode, Kerala, India

N. S. Pradeep, L. K. Edison (eds.), *Microbial Beta Glucanases*, Interdisciplinary Biotechnological Advances, https://doi.org/10.1007/978-981-19-6466-4_10

10.1 Introduction

Currently, industrial enzymes are merchandized as commodity products, which tremendously require continuous production and moderate manufacturing cost with the intention of persisting as economical in the global markets. Recombinant enzyme production through heterologous expression is considered a pivotal pace for establishing cost-effective, unconventional productions in modern industrial biotechnology. *E. coli* was chosen for aiding in heterologous protein production as it allowed rapid growth, facile management, cost-effective fermentation and well-comprehend genetics (Fath et al. 2011). However, species-specific discrepancies in codon usage due to genetic code degeneracy are usually cited as a major hindrance that adversely impacts protein expression levels in *E. coli* (Zhou et al. 2004). The occurrence of rare codons can diminish the translation rate and also generate translation errors, which are interrelated with low levels of cognate tRNAs species in the cell (Ikemura 1981).

The usage of synonymous codons without disturbing the original protein sequence is called codon optimization. This method greatly enhances overall gene expression and heterologous proteins, this clarifies the vast dissimilarity between the codon usage of the host cell and the inherent heterologous protein-encoding sequence (Yu et al. 2013). To date, a higher number of codon-optimized and synthesized genes have demonstrated successful overexpression (Jung and McDonald 2011). Codon-optimized genes mimic the codon bias of host cells or eliminate rare codons thus substantially increasing the translation rates with great accuracy. With the rapid progress of these approaches, numerous online tools for codon optimization have been developed (Lanza et al. 2014), which also assisted to remove the gene expression inhibitory elements such as certain DNA/mRNA-based bad motifs and RNA secondary structures (Graf et al. 2004).

Streptomyces genome encompasses elevated (>70%) GC content and different compositional bias in codon usage might significantly decrease the protein expression (Liu et al. 2016) in a heterologous host. Moreover, significant codon usage heterogeneity was found between different *Streptomyces* genomes (Wu et al. 2005). Hence, the production of proteins in *E. coli* as a host system produces low yield expression because of these codon adaptation discrepancies. Previous findings strongly advised that a comprehensive codons optimization strategy with concurrent modulation of multiparameters such as GC content, codon quality, ribosome entry site, and secondary structure formation might be the pre-eminent solution for overexpression in a heterologous system (Xiong et al. 2006; Teng et al. 2007). This optimization of the target protein-coding sequence as per the demands of the host leads to a protein production yield of 10- to 50-fold (Sinclair and Choy 2002; Kanekiyo et al. 2005).

10.2 Strategies for Heterologous Production

The production of recombinant enzymes has a notable demand commercial market due to their dramatically increasing application in industrial sectors. Heterologous expression is a suitable way to produce large- or small-scale production of enzymes. The obtained genetic and biochemical knowledge of proteins enables economically feasible large-scale production. Production of recombinant beta-glucanases has been achieved in a variety of bacterial expression systems such as *Escherichia coli* and *Bacillus* (Zeeshan et al. 2018; Lambertz et al. 2014; Boisramé and Gaillardin 2009). However, full-length proteins with significant enzyme activity and high yield look for many challenges such as lack of protease resistance, thermostability, misfolding, cellular stresses, and inefficiency in extracellular secretion. Design of expression vector, gene dosage, regulation of transcription and translation, host design, codon usage, and fermentation conditions are the various strategies for accomplishing high-level heterologous production of recombinant proteins at low cost (Fig. 10.1).

To ensure optimal expression, numerous factors were taken into consideration while designing the expression plasmid. Features of promoters such as position, strength and its control by a regulatory gene are the essential factors to be considered. An advantageous promoter shows a strong and tightly regulated expression level, simple induction, effortlessly transferable feature to other *E. coli* strains, the

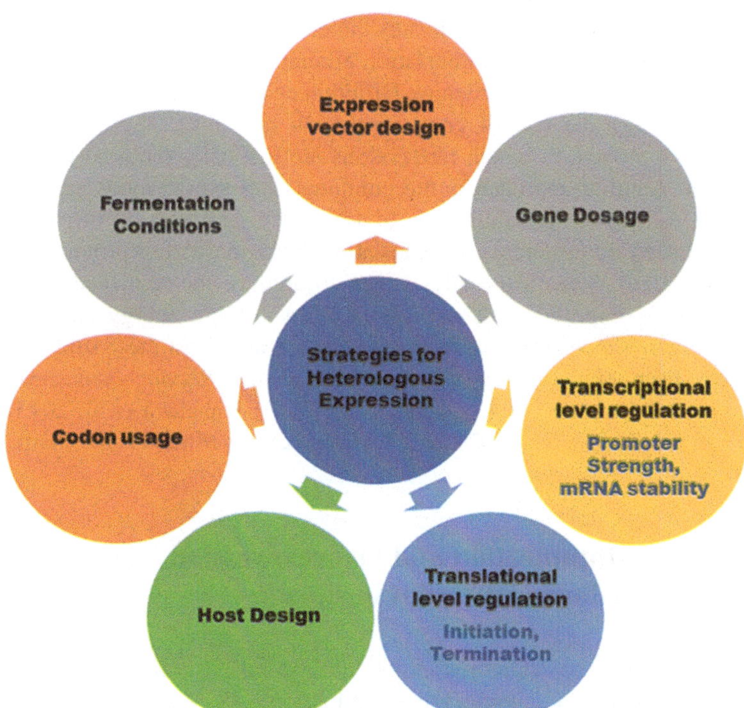

Fig. 10.1 General strategies for heterologous protein production

distance between promoter and ribosome binding site (RBS), antibiotic resistance, and cost-effectiveness (Sprengart and Porter 1997; Ringquist et al. 1992). mRNA stability is determined by the distance between Shine–Dalgarno (SD) site in vector and start codon. The sequence between these regions should eliminate the secondary structure of mRNA, it further enhances translation efficiency. The transcription termination sequence located the downstream of expression vector protects the mRNA from exonucleolytic degradation by forming a stem–loop structure. The second strategy is the gene dosage that refers to the copy number. It determines the plasmid stability and also helps to quantify the gene expression. The maximum gene expression is generally anticipated by high copy number plasmids.

10.3 Codon Optimization for Recombinant Expression

One of the fundamental technologies for efficient protein production in bacteria is recombinant protein expression. The expression level of recombinant protein can be increased by optimizing the protein-coding sequence by synonymous substitution, called codon optimization (Gustafsson et al. 2004). Optimized codon expression helps the proper folding of the synthesized protein as well as increases the functional activity. Gene coding sequences are optimized by considering sequence features that affect protein expression levels (Gustafsson et al. 2012; Quax et al. 2015; Parret et al. 2016; Brule and Grayhack 2017). The key factors considered for codon optimization include transcription efficiency, translation efficiency, protein folding, tRNA abundance, and mRNA secondary structure. Different codon optimization approaches are used for increasing the rate of gene expression in a host organism. One is the conventional approach, in which rare codons are manually replaced by frequently used codons. Another approach is the computational synonymous substitution of codons that leads to the destabilization of secondary structures of mRNA. Destabilization enhances the efficiency of translation, thus increasing protein expression (Saito et al. 2019). Statistical analysis uses to analyze the sequence features that influence protein expression. It provides mRNA folding energy and codon frequency. *Escherichia coli* remains the model organism to host these studies. Though some have been conducted with a rare hosts like *Rhodococcus erythropolis* (Nakashima and Tamura 2004; Kasuga et al. 2017; Kitagawa et al. 2013, 2018), *Pichia pastoris* (Teng et al. 2007), and *Bacillus subtilis* (Shields and Sharp 1987).

10.4 Codon Optimization and Overexpression of Beta-Glucanase Genes in *E. coli*

Current knowledge in sequence manipulation as per host demands without losing the integrity of the final protein product, in combination with de novo synthesis of genes results in an efficient recombinant protein expression that meets the industrial level demands for high-scale production. Codon optimization was deliberated as an efficient technique to improve the level of heterologous gene expression of beta-glucanase genes, while the codon alteration through traditional choices like site-directed mutagenesis could yield mRNA instability sequentially reduced the protein yield (Edison et al. 2020; Lammertyn et al. 1996). With the development of advanced algorithms, the traditional sequence optimization strategies have been replaced by multiple measures to calculate ideal solutions for suitable experimental requirements. The calculation and merging of multiple optimization parameters such as codon usage, GC content, restriction sites, ribosomal entry sites, RNA instability motifs, and hairpin loops have been considered for the better expression of beta-glucanase genes from *Streptomyces* sp. in the *E. coli* host. The calculations and optimizations can be achieved by performing a comprehensive search absolutely in a small sequence window, which is progressed along with the entire reading frame (Edison et al. 2020; Fath et al. 2011). The overall process of codon manipulation and heterologous gene expression of beta-glucanase genes is explained in Fig. 10.1.

The codon optimization of exo-β-1,4-glucanase (exo14) from *Streptomyces althioticus* TBG-MR17 and endo-β-1,3-glucanase (endo13) from *Streptomyces cinereoruber* sub sp. *Cinereoruber* TBG-AL13 has been done using GeneArt™ GeneOptimizer® expert software from Life Technologies-Invitrogen for the effective overexpression of proteins in heterologous host *E. coli* (Edison et al. 2020). The software employs a multiparameter deterministic algorithm that generates a sliding window technique for the optimization process. This algorithm runs the sequences as unidirectional like naturally occurring inside the cell, starting from 5′ and ending at 3′. It helps to remove non-appropriate motifs such as GC-rich sequences, internal ribosomal entry sites, RNA instability motifs, repeat sequences, and RNA secondary structures. After optimization, the genes synthesize de novo by oligonucleotide assembly and cloned into pMA-T cloning vectors, and transforms into *E. coli* K12 OmniMAX™ 2 T1R cells. Plasmid's purification and confirmation of the presence of the gene of interest is done through PCR amplification and sequencing to confirm the sequence identity of the final construct (Fig. 10.2). GeneOptimizer algorithm takes into consideration nearly 50 parameters to evaluate the optimal gene sequence for elevated expression while preserving the final protein integrity (Raab et al. 2010).

To improve the protein expression, the differences between relative codon adaptiveness were compared to that of *E. coli* and *Streptomyces* spp. The GeneOptimizer® software adapted codon usage to meet *E. coli* codon bias requirements. A total of 431 codons from 578 total codons of the enzyme exo-β-1,4-glucanase wild sequence (72% of changes) was elaborated, and the codon adaptation index (CAI) shifted from 0.73 (wild-type exo-β-1,4-glucanase) to 0.97 (optimized exo-β-1,4-glucanase). The GC content decreased from 69 to 49%. Sequences of wild type and optimized exo-β-1,4-glucanase with a corresponding amino acid sequence are shown in Fig. 10.3. Codon optimization of endo-β-1,3-glucanase gene by GeneOptimizer® manipulated 324 codons from 430 codons (75%) of the wild type. The CAI index moved from 0.67 to 0.98 and GC content showed a reduction from 71 to 53%. The wild type and codon calibrated endo-β-1,3-glucanase sequences with its associated amino acid sequence are detailed

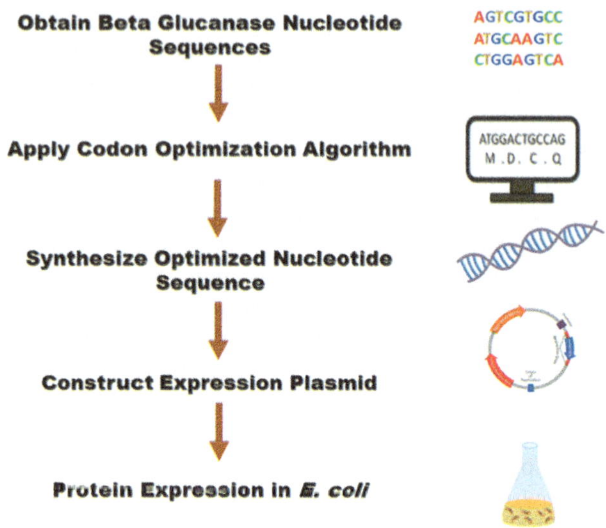

Fig. 10.2 The overall process of codon optimization and heterologous expression of beta-glucanase genes

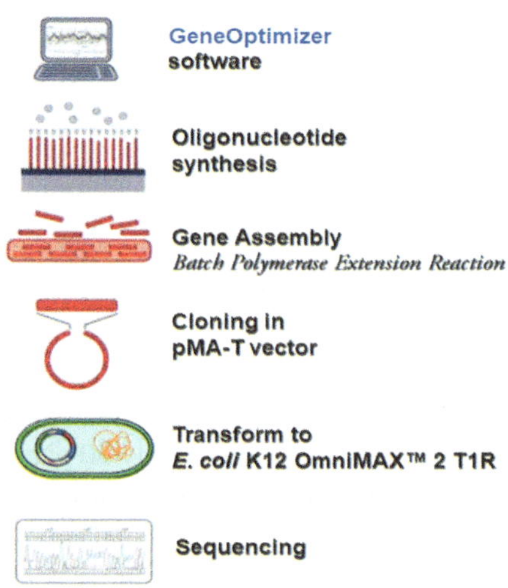

Fig. 10.3 Schematic overview of GeneArt™ codon optimization

in Fig. 10.4. The newly obtained CAI value and GC content resulted in increased mRNA half-life and elevated expression. High GC content prevailing in gene of interest, in a background of *E. coli* codon bias is the contributing factor to low protein expression in the *E. coli* host (Redwan 2006). In order to efficiently express proteins in the *E. coli* host, it was recommended to change the GC content of both gene sequences (70%) to match the requirements of the *E. coli* genome GC content (Edison et al. 2020).

Fig. 10.4 Wild type and optimized exo-β-1,4-glucanase sequence with the corresponding amino acid sequence

Champion™ pET101 directional TOPO® Expression vector (5753 bp) was employed for the expression of wild and optimized beta-glucanase genes due to their highly successful production of recombinant proteins (Edison et al. 2020; Tokano et al. 2008; Zarif et al. 2013; Kishore et al. 2015) and topoisomerase helped in ligation of blunt-end PCR product without ligase (Shuman 1994). Full length of wild (wlEXO14 and wlENDO13) and optimized (optEXO14 and optENDO13) beta-glucanase sequences have been amplified precisely for cloning in pET101 directional TOPO® cloning vector. pET101 is a linear vector that carries 5′-CCCTT-3′ sequences at two ends. The free 3'phosphate group of each 3'thymidine is covalently bound with enzyme topoisomerase I. When the vector is mixed with the PCR product, the covalently linked topoisomerase I assisted to ligate the compatible 5′-OH end of the PCR product (because primers are designed without adding the 5′-P) to the 3'end of vector strand, and the enzyme is released. Additionally, the GTGG- 3′overhang sequence in the vector annealed with the 5′-CACC overhang of the PCR product, which confirms the insert was correctly oriented before being ligated by topoisomerase. Schematic representation showing the detailed architecture of recombinant expression constructs in Fig. 10.5. Manipulated codon of synthetic human insulin gene cloned in pET101/D-TOPO expression vector expressed in *E. coli* cells produced a high yield of proinsulin protein (El-Aziz et al. 2014). Habib-ur-Rehman et al. (2018) confirmed that *E. coli* efficiently expressed recombinant pullulanase gene of *Pyrobaculum calidifontis* origin, after being cloned in pET101/D-TOPO vector, after codon manipulation. The induction temperature is one of the critical factors that influence extremely on protein expression (Salazar et al. 2001).

The resulting overexpression was evaluated in terms of relative fold expression of optimized proteins as per the percentage peak area of intensity plots of each band measured with ImageJ software (Fig. 10.6). ImageJ can compare the density of each band in SDS-PAGE gel, by making comparisons between bands in different lanes (Motojima 2018). The modified exo-1,4-glucanase gene showed an elevation of 13-fold protein expression in comparison to wild codons, whereas the optimized endo-1,3-glucanase gene displayed a 3.75-fold expression in comparison to wild genes. Sequence optimization in the above two enzymes resulted in hiked expression compared to native ones. Table 10.1 shows the comparative data of CAI, GC content, and expression profile of wild and optimized genes. The Study by Yadava and Ockenhouse (2003) revealed that malarial vaccine production in host *E. coli* improved vastly after codon optimization, when compared with production by original sequence. Commercial level production was satisfied in codon-optimized human cystatin gene with *E. coli* as host (Qing et al. 2012) (Fig. 10.7).

The comparative study of the exo-β-1,4-glucanase and endo-β-1,3-glucanase enzyme activities and protein concentrations showed the enzyme activities of optexo14 and optendo13 showed an increase of 16.56- and 6.32-folds, respectively. The improved function of proteins is due to the production of a significantly high yield with a more native folded state, which can be established by the codon optimization approach (Li et al. 2014). Previous reports also proved that codon optimization also led to improved catalytic activity in enzymes, when compared to

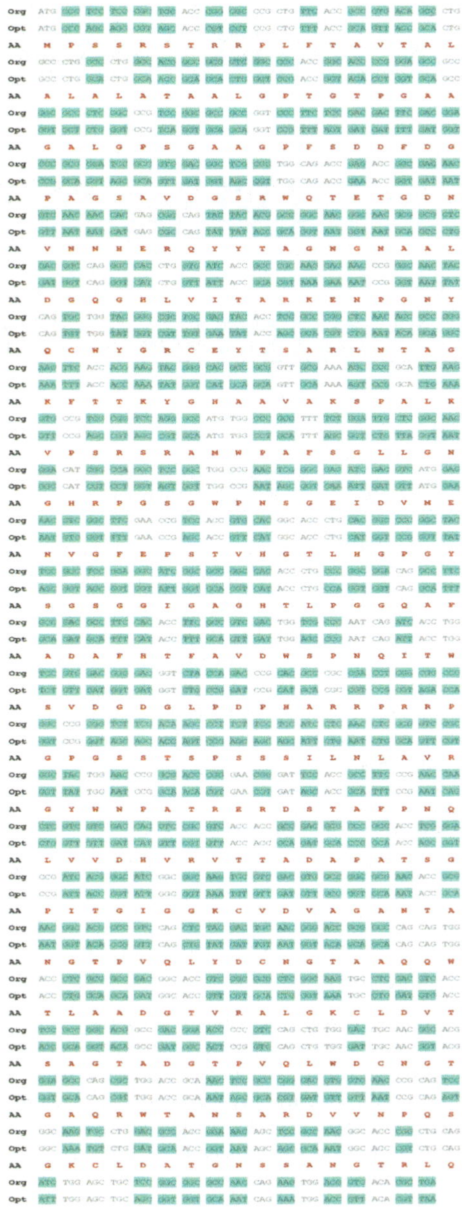

Fig. 10.5 Wild type and optimized endo-β-1,3-glucanase sequence with the corresponding amino acid sequence

native ones. *α*-Amylase gene isolated from *Bacillus licheniformis* was subjected to codon optimization, resulting in 2.62-fold higher expression in the host *Pichia pastoris* (Wang et al. 2015). The expression of lipase from *Candida rugosa* in *P. pastoris* was enhanced by 4.6-fold expression by optimizing its codons to favour the codon bias of *P. pastoris* (Chang et al. 2006). The wild-type lipase gene from

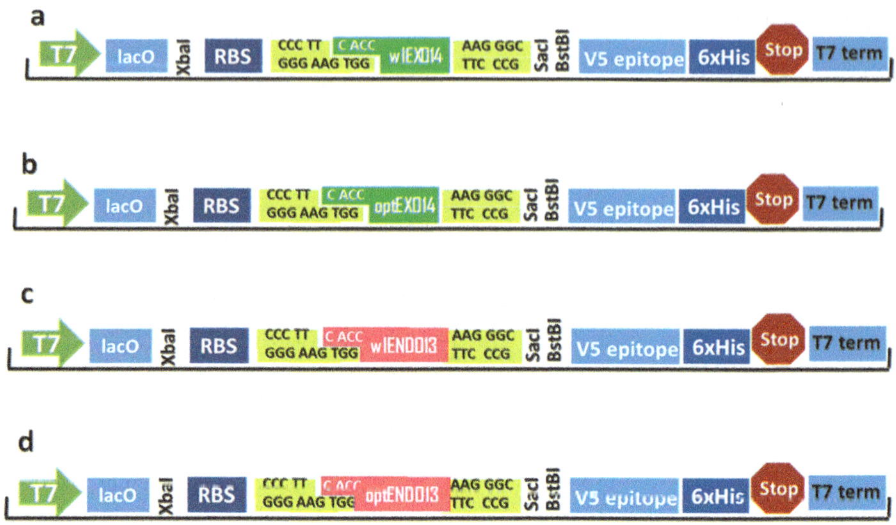

Fig. 10.6 Schematic representation of the detailed architecture of pET101 recombinant expression constructs of wild-type *Streptomyces* (wlEXO14 and wlENDO13) and optimized beta-glucanase (optEXO14 and optENDO13) genes for the heterologous expression in *E. coli* (**a**) pET101/wlEXO14 construct. (**b**) pET101/optEXO14 construct. (**c**) pET101/wlENDO13 construct. (**d**) pET101/optENDO13 construct. All the recombinant constructs worked under the T7 promoter. CACC overhang of the PCR product is captured by the complementary GTGG overhang sequence in the vector. C-terminal V5 epitope (14 amino acid epitope) is specific to AntiV5 antibody helped for probing the expression product. 6xHis tag permitted the expressed proteins for the Ni-affinity purification

Table 10.1 Comparison of CAI, GC content, and expression profile of wild and optimized genes

		CAI[a]		GC%		Codon	Codon		
Protein	Length	wl	opt	wl	opt	Altered (No)	Altered (%)	Ratio of expression[b]	Expression statistics
exo14	578	0.73	0.97	69	49	431	74	1:13	+
endo13	430	0.67	0.98	71	53	324	75	1:3.75	+

[a]CAI-codon adaptation index is a measure of the relative adaptiveness values of the used codons which represents the ratio of the frequency of codons in codon usage of an expression system and frequency of synonymous codons for the same amino acid, which gives a value of 1 for optimal codon and less frequently used codons are scaled down accordingly

[b]Ratio of expression calculated by analyzing the intensity of each band as the percentage peak area of intensity plots using ImageJ (NIH, USA) gel quantification through densitometry

Candida rugosa produced no lipase activity in heterologous host *P. pastoris*, while its optimized gene product produced the hydrolytic activity of 4.7 $U.mL^{-1}$ in *P. pastoris* (Xu et al. 2010). The total protein content of optexo14 and optendo13 revealed 12.96-fold and 5.19-fold increases, respectively. Nearly, similar fold increases in the protein quantities were detected when compared with densitometry determination of fold expressions of both exo14 and endo13 proteins. Codon-optimized genes flash a positive impact on expression at different molecular

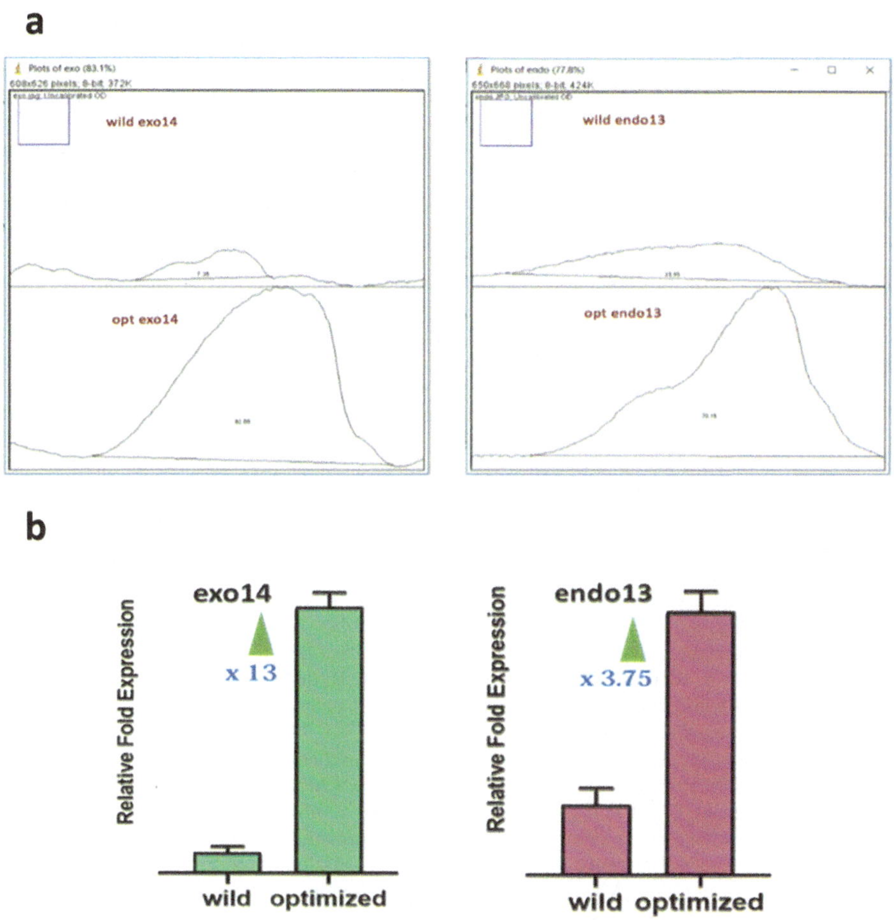

Fig. 10.7 Overexpression of wild and optimized exo-β-1,4-glucanase (exo14) and endo-β-1,3-glucanase proteins (endo13). (**a**) The intensity plot of overexpressed bands was determined by densitometry, using ImageJ software (**c**) The relative fold expression levels of overexpressed proteins were calculated based on the percentage peak area of each intensity plot

processes such as transcription and translation and also elevate mRNA stability. These effects, all together contribute to protein yield and activity of recombinant proteins (Fath et al. 2011).

Codon optimization increases gene expression at certain levels. Codon optimization and production of proteins in a heterologous system is a powerful tool to produce recombinant products. Currently, researchers are focusing to reduce the rate of gene synthesizing technology by assembling the DNA fragments. In future, more developments in bioinformatic tools may provide additional assistance towards synthetic biology.

References

Boisramé A, Gaillardin C (2009) Heterologous expression and characterization of a beta-1,6-glucanase from aspergillus fumigatus. Appl Microbiol Biotechnol 82(4):663–669

Brule CE, Grayhack EJ (2017) Synonymous codons: choose wisely for expression. Trends Genet 33(4):283–297

Chang SW, Lee GC, Shaw JF (2006) Codon optimization of Candida rugosa lip1 gene for improving expression in Pichia pastoris and biochemical characterization of the purified recombinant LIP1 lipase. J Agric Food Chem 54(3):815–822

Edison LK, Dan VM, Reji S (2020) A strategic production improvement of Streptomyces Beta glucanase enzymes with aid of codon optimization and heterologous expression. R. and Pradeep, N.S. Biosci Biotechnol Res Asia 17(3):587–599

El-Aziz GA, Abdel-Aziz SH, Reda MM, Baehen M, Baeshen NA, Redwan EM (2014) Synthesis of human insulin gene in vitro through computational methodology. Life Sci J 11(1):27–34

Fath S, Bauer AP, Liss M, Spriestersbach A, Maertens B, Hahn P, Ludwig C, Schäfer F, Graf M et al (2011) Multiparameter RNA and codon optimization: a standardized tool to assess and enhance autologous mammalian gene expression. PLoS One 6(3):e17596

Graf M, Deml L, Wagner R (2004) Codon-optimized genes that enable increased heterologous expression in mammalian cells and elicit efficient immune responses in mice after vaccination of naked DNA. Methods Mol Med 94:197–210

Gustafsson C, Govindarajan S, Minshull J (2004) Codon bias and heterologous protein expression. Trends Biotechnol 22(7):346–353

Gustafsson C, Minshull J, Govindarajan S, Ness J, Villalobos A, Welch M (2012) Engineering genes for predictable protein expression. Protein Expr Purif 83(1):37–46

Habib-Ur-Rehman S, Siddiqui MA, Qayyum A, Bano A, Rashid N (2018) Gene expression in Escherichia coli and purification of recombinant type II pullulanase from a hyperthermophilic archaeon, Pyrobaculum calidifontis. Pak J Zool 50(4):1381–1386

Ikemura T (1981) Correlation between the abundance of Escherichia coli transfer RNAs and the occurrence of the respective codons in its protein genes. J Mol Biol 146(1):1–21

Jung SK, McDonald K (2011) Visual gene developer: a fully programmable bioinformatics software for synthetic gene optimization. BMC Bioinformatics 12:340

Kanekiyo M, Matsuo K, Hamatake M, Hamano T, Ohsu T, Matsumoto S, Yamada T, Yamazaki S, Hasegawa A et al (2005) Mycobacterial codon optimization enhances antigen expression and virus-specific immune responses in recombinant Mycobacterium bovis bacille Calmette-Guerin expressing human immunodeficiency virus type 1 gag. J Virol 79(14):8716–8723

Kasuga K, Sasaki A, Matsuo T, Yamamoto C, Minato Y, Kuwahara N, Fujii C, Kobayashi M, Agematu H et al (2017) Heterologous production of kasugamycin, an aminoglycoside antibiotic from Streptomyces kasugaensis, in Streptomyces lividans and Rhodococcus erythropolis L-88 by constitutive expression of the biosynthetic gene cluster. Appl Microbiol Biotechnol 101(10): 4259–4268

Kishore V, Nishita KP, Manonmani HK (2015) Cloning, expression and characterization of L-asparaginase from pseudomonas fluorescensfor large scale production in E. coli BL21. Biotechnology 5:975.e81

Kitagawa W, Mitsuhashi S, Hata M, Tamura T (2018) Identification of a novel bacteriocin-like protein and structural gene from Rhodococcus erythropolis JCM 2895, using suppression-subtractive hybridization. J Antibiot (Tokyo) 71(10):872–879

Kitagawa W, Ozaki T, Nishioka T, Yasutake Y, Hata M, Nishiyama M, Kuzuyama T, Tamura T (2013) Cloning and heterologous expression of the aurachin RE biosynthesis gene cluster afford a new cytochrome P450 for quinoline N-hydroxylation. Chembiochem 14(9):1085–1093

Lambertz C, Garvey M, Klinger J, Heesel D, Klose H, Fischer R, Commandeur U (2014) Challenges and advances in the heterologous expression of cellulolytic enzymes: a review. Biotechnol Biofuels 7(1):135

Lammertyn E, Van Mellaert L, Bijnens AP, Joris B, Anné J (1996) Codon adjustment to maximise heterologous gene expression in Streptomyces lividans can lead to decreased mRNA stability and protein yield. Mol Gen Genet 250(2):223–229

Lanza AM, Curran KA, Rey LG, Alper HS (2014) A condition-specific codon optimization approach for improved heterologous gene expression in Saccharomyces cerevisiae. BMC Syst Biol 8:33

Li YM, Li DJ, Xu XJ, Cui M, Zhen HH, Wang Q (2014) Effect of codon optimization on expression levels of human cystatin C in Pichia pastoris. Genet Mol Res 13(3):4990–5000

Liu S, Wang M, Du G, Chen J (2016) Improving the active expression of transglutaminase in Streptomyces lividans by promoter engineering and codon optimization. BMC Biotechnol 16(1):75

Motojima F (2018) Rapid electrophoretic staining and destaining of polyacrylamide gels. Methods Protoc 1(2):13

Nakashima N, Tamura T (2004) Isolation and characterization of a rolling-circle-type plasmid from Rhodococcus erythropolis and application of the plasmid to multiple-recombinant-protein expression. Appl Environ Microbiol 70(9):5557–5568

Parret AH, Besir H, Meijers R (2016) Critical reflections on synthetic gene design for recombinant protein expression. Curr Opin Struct Biol 38:155–162

Qing W, Mei C, Zhen H, Zhu J (2012) Codon preference optimization increases prokaryotic cystatin C expression. J Biomed Biotechnol 732017:1–7

Quax TE, Claassens NJ, Söll D, van der Oost J (2015) Codon bias as a means to fine-tune gene expression. Mol Cell 59(2):149–161

Raab D, Graf M, Notka F, Schödl T, Wagner R (2010) The GeneOptimizer algorithm: using a sliding window approach to cope with the vast sequence space in multiparameter DNA sequence optimization. Syst Synth Biol 4(3):215–225

Redwan EL (2006) The optimal gene sequence for optimal protein expression in Escherichia coli: principle requirements. Arab J Biotechnol 9:493–510

Ringquist S, Shinedling S, Barrick D, Green L, Binkley J, Stormo GD, Gold L (1992) Translation initiation in Escherichia coli: sequences within the ribosome-binding site. Mol Microbiol 6(9): 1219–1229

Saito Y, Kitagawa W, Kumagai T, Tajima N, Nishimiya Y, Tamano K, Yasutake Y, Tamura T, Kameda T (2019) Developing a codon optimization method for improved expression of recombinant proteins in Actinobacteria. Sci Rep 9(1):8338

Salazar O, Molitor J, Lienqueo ME, Asenjo JA (2001) Overproduction, purification, and characterization of β-1,3-vlucanase type II in Escherichia coli. Protein Expr Purif 23(2):219–225

Shields DC, Sharp PM (1987) Synonymous codon usage in Bacillus subtilis reflects both translational selection and mutational biases. Nucleic Acids Res 15(19):8023–8040

Shuman S (1994) Novel approach to molecular cloning and polynucleotide synthesis using vaccinia DNA topoisomerase. J Biol Chem 269(51):32678–32684

Sinclair G, Choy FYM (2002) Synonymous codon usage bias and the expression of human glucocerebrosidase in the methylotrophic yeast Pichia pastoris. Protein Expr Purif 26(1):96–105

Sprengart ML, Porter AG (1997) Functional importance of RNA interactions in selection of translation initiation codons. Mol Microbiol 24(1):19–28

Teng D, Fan Y, Yang YL, Tian ZG, Luo J, Wang JH (2007) Codon optimization of bacillus licheniformis −1,3-1,4- glucanase gene and its expression in Pichia pastoris. Appl Microbiol Biotechnol 74(5):1074–1083

Tokano DV, Kawaichi ME, Venancio EJ, Vidotto MC (2008) Cloning and characterization of the iron uptake gene IutA from avian Escherichia coli. Braz Arch Biol Technol 51(3):473–482

Wang JR, Li YY, Liu DN, Liu JS, Li P, Chen LZ, Xu SD (2015) Codon optimization significantly improves the expression level of α -amylase gene from bacillus licheniformis in Pichia pastoris. Biomed Res Int 2015:248680

Wu G, Culley DE, Zhang W (2005) Predicted highly expressed genes in the genomes of Streptomyces coelicolor and Streptomyces avermitilis and the implications for their metabolism. Microbiology (Reading) 151(7):2175–2187

Xiong AS, Yao QH, Peng RH, Zhang Z, Xu F, Liu JG, Han PL, Chen JM (2006) High level expression of a synthetic gene encoding Peniophora lycii phytase in methylotrophic yeast Pichia pastoris. Appl Microbiol Biotechnol 72(5):1039–1047

Xu L, Jiang X, Yang J, Liu Y, Yan Y (2010) Cloning of a novel lipase gene, lipJ08, from Candida rugosa and expression in Pichia pastoris by codon optimization. Biotechnol Lett 32(2):269–276

Yadava A, Ockenhouse CF (2003) Effect of codon optimization on expression levels of a functionally folded malaria vaccine candidate in prokaryotic and eukaryotic expression systems. Infect Immun 71(9):4961–4969

Yu P, Yan Y, Gu Q, Wang X (2013) Codon optimisation improves the expression of Trichoderma viride sp. endochitinase in Pichia pastoris. Sci Rep 3:3043, 3

Zarif R, Sankian M, Gholubi A, Farshadzadeh Z, Soleimanpour S, Youssefi F, Abdolreza V (2013) Cloning and expression of Mycobacterium tuberculosis major secreted protein antigen 85B (Ag85B) in Escherichia coli. Jundishapur J Microbiol 6(2):112–116

Zeeshan N, Naz S, Afroz A, Zahur M, Zia S (2018) Heterologous expression and enhanced production of β-1,4-glucanase of Bacillus halodurans C-125 in Escherichia coli. Electron J Biotechnol 34:29–36

Zhou Z, Schnake P, Xiao L, Lal AA (2004) Enhanced expression of a recombinant malaria candidate vaccine in Escherichia coli by codon optimization. Protein Expr Purif 34(1):87–94

Chapter 11
Industrial Production and Purification of Recombinant Beta-Glucanases

Lekshmi K. Edison, P. K. Satheeshkumar, and N. S. Pradeep

Abstract Industrial sector is always in search of enzymes with special properties. The enzymatic conversion of cereal beta-glucans by beta-glucanases is a promising approach for preventing the negative effects of beta-glucans in the industrial processes of brewing as well as animal feed industries. Complete degradation of beta-glucans requires the synergic action of different beta-glucanases. Hence, enzyme cocktails are preferably suggested for the industrial applications. Many technical and economic uncertainties hinder the cost-effective production of industrially important beta-glucanases. The cost of enzyme production and purification is significant challenge during industrial production. The application of genetic engineering strategies is a potential way to increase volumetric productivity and reduce production costs. Recombinant enzyme production using prokaryotic expression systems is a low-cost production strategy using high cell-density cultures for obtaining a high volume of an enzyme. This chapter discusses the techno-economic evaluation in the production, purification and characterization of recombinant beta-glucanases.

Keywords Recovery · Production media · Substrate specificity · Cocktail mix

11.1 Introduction

Enzyme technology, an interdisciplinary area, comprises identification, production and purification of more competent and beneficial enzymes for the ultimate advantage of humanity. Industrial enzymes signify the core of biotechnology processes. Nowadays, foremost research and development initiatives look not only enzyme

L. K. Edison
Rajiv Gandhi Centre for Biotechnology (RGCB), Thiruvananthapuram, Kerala, India

P. K. Satheeshkumar
Centre of Advanced Study in Botany, Institute of Science, Banaras Hindu University, Varanasi, India

N. S. Pradeep (✉)
KSCSTE-Malabar Botanical Garden and Institute for Plant Sciences, Kozhikode, Kerala, India

N. S. Pradeep, L. K. Edison (eds.), *Microbial Beta Glucanases*, Interdisciplinary Biotechnological Advances, https://doi.org/10.1007/978-981-19-6466-4_11

production but also for the improvement of process development. All enzymes are selective for particular substrates. The production of enzymes that are marketable and economical is the major focus of research in the biotechnology industry. Diverse scientific areas like molecular biology, biochemistry and protein engineering contributed to the progress of enzyme technology. The first industrially produced microbial enzyme was taka-diastase, an amylase, extracted from fungi in 1896. It was used to cure digestive disorders (Oshima and Itaya 1926). Microorganisms are the greatest convenient and economically important enzyme source, and produce copious quantities under apt growth conditions. A plethora of microorganisms is used with low-cost media for the large-scale production of industrial enzymes. Enzyme production can be increased easily by the genetic manipulation of microbial systems through genetic engineering techniques. In addition, purification and final recovery of enzymes are easier with microbial systems.

The advancements in recombinant DNA technology, as well as protein engineering, revolutionized the production of proteins in massive quantities. Although it has been produced by using modern technologies, it requires subsequent purification to remove all the contaminants that create complications in solubility, structural integrity and functional properties. The genetic engineering approach makes protein expression and purification relatively simple and powerful by making fusion proteins which permits the purification by affinity techniques without having any prior knowledge about the properties of targeted proteins (Scopes 1995). Characterization of newly developed products has an important space in today's science. Once a pure protein is acquired, it is essential for determining the detailed physio-chemical characteristics as well as assay conditions on laboratory scale for achieving maximum activity, stability and shelf life for industrial applications. Here, the inclusive aim is to launch a link between biological activities and external parameters like pH and temperature. The study of catalytic switching characteristics such as temperature and pH optima, stability under different pH and temperatures, substrate specificity, effects of cofactors and inhibitors and kinetic parameters is crucial to use the catalytic proteins in respective industrial sectors.

Beta-glucanases are glucanolytic enzymes that depolymerize beta-glucan molecules found in the cell wall of plants, fungi and algae leading to the release of D-glucose molecules, which serve as carbon source (Stahmann et al. 1992; Shao-jun et al. 2001; Tang-Yao 2002). Beta-glucanases are industrially important enzymes with commercial applications in breweries and animal feed industries for the degradation of cereal beta-glucans. It can be used as an individual enzyme or cocktail mix of different beta-glucanases or with other enzymes (Bajomo and Young 1992). It is also used as an additive in laundries and used for saccharification. Industrial application of beta-glucanases has great attention all over the world (Srinivasa 2005). Commercial beta-glucanases are widely produced from microorganisms like *Bacillus subtilis, Streptomyces* sp., *Pseudomonas fluorescence, Disporotrichum dimorphosphorum, Aspergillus niger and Penicillium species*. Beta-glucanases from microbial sources show high thermostability and substrate affinity (Aono et al. 1995). Due to the high potential and wide applications of beta-glucanases in the world market, the research and development of many industries are trying to introduce new strategies for the large-scale production, purification and characterization.

11.2 Industrial Production Beta-Glucanases

Industrial production of beta-glucanases is done mainly through two different methods namely (1) production using lab-adapted natural strains and (2) by heterologous expression using any expression hosts engineered for industrial-scale recombinant protein expression. In both cases once the initial standardizations are done for the maximum expression of the protein with desirable properties, the commercial production is mainly achieved through submerged fermentation or solid-state fermentation, in huge bioreactors called fermenters. The downstream processing depends on the purity required and further depends on the application of enzymes. Large volume beta glucanases that are generated primarily for use in bioremediation or the detergent sector are not usually purified. However, beta-glucanases produced for the brewing industry, animal feed industry, or other food industries are required to be highly pure in nature. Industrial enzyme production is divided into different phases including the identification of organisms, formulation culture media, production process, recovery and purification (Fig. 11.1).

The search for the microbial strains with the potential to be exploited in the industrial production of beta-glucanase normally starts with the places where the plant origin waste materials are dumped. Conventional microbiology procedures like sample collection, screening, raising of pure cultures and biochemical tests for enzyme identification will be done to identify a few microbial strains which express beta-glucanase. Among the strains identified at the taxonomic levels, the biochemical tests will be used to find out the best strain in terms of quantity and quality of the enzyme produced. Lab-scale small volume cultures are used to study the rate of enzyme production, the recovery and purification strategies to be developed and physico-biochemical and biological properties of the enzyme. The microbial strain which produces a maximum quantity of enzymes with good/excellent functional properties within a short period is the major criteria considered for organism selection. Other important factors considered for the selection are substrate affinity and specificity, rate of reaction, temperature stability, pH stability and inhibitor sensitivity. Strain improvement using site-directed mutagenesis can positively influence the enzyme production. Mostly, the preferable microbial strain is an extracellular enzyme producer, because of the ease of recovery and purification of the target protein from the media. Beta-glucanase-producing host strains, for applications in breweries and animal feeds should have GRAS (generally recognized as safe) status (Burton et al. 2002; Jaeger et al. 2001; Powell et al. 2001; Farinas et al. 2001). All the nutrients in the enzyme production medium should ensure the adequate growth of microbial strains which eventually increases the enzyme quantities. Nutritionally safe, low-cost media components are generally used. Commonly used substrates are corn steep liquor, yeast extract, soybean meal, molasses, whey and starch hydrolysate.

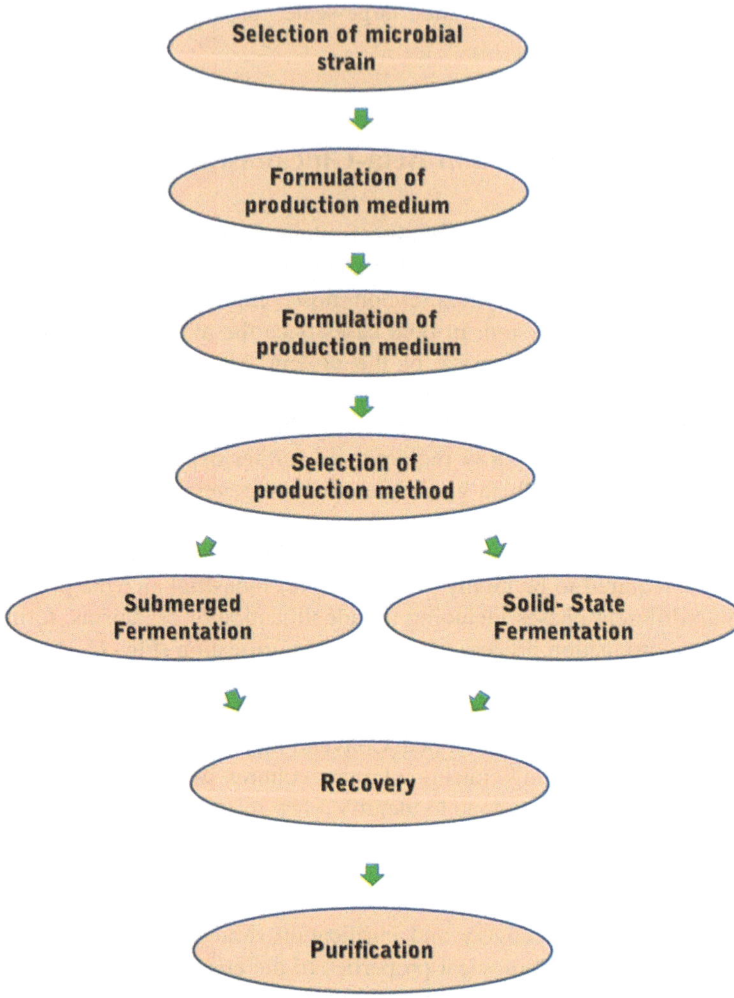

Fig. 11.1 Steps involved in industrial enzyme production

Once the lab-scale research proves the efficacy of a strain to be developed as an industrial producer of beta-glucanase, the next step is to check its adaptability to grow and produce the target protein in large volume cultures under aseptic conditions. The extensively used industrial enzyme production is through submerged fermentation and solid-state fermentation. The optimization of media composition, inoculum and culture conditions are necessary for attaining the best result on production. Major industrial production is done through submerged fermentation using well-equipped bioreactors. In submerged fermentation, the yields are high and very fewer chances of infection. A liquid medium is used for submerged fermentation. Oxygen supply is crucial in submerged reactors. Different types of submerged fermentations are batch culture, perfusion batch culture, fed-batch culture and continuous culture. In batch culture, fermentation is done as batches using the fixed volume of the medium. In perfusion batch culture, withdrawal and addition of an equal volume of culture and medium are performed at a definite interval. In fed-batch culture, the nutrients are fed during the batch culture. In continuous culture, withdrawal of the media with the product and corresponding addition of fresh medium is done during the log phase of microbial growth. Batch fermentation is primely used for enzyme production. Ultrafiltration or gel filtration is used for the recovery of enzymes after cell removal. The concentrated enzyme is then mixed with preservatives and drying. The submerged fermentation is uneconomical due to the high cost. Due to the possibility of using agro-industrial wastes as a substrate, solid-state fermentation is now receiving increased interest (Pandey 2003). So the solid-state fermentation is attractive to reduce the production cost of enzymes. High productivity, high product stability, low-cost production, low catabolic repression, use of mixed microbial strains and use of water-insoluble substrates are the advantages of solid-state fermentation (Hölker et al. 2004). However, scale-up, biomass estimation and purification are major challenges of solid-state fermentation. Further, the recovery and purification are also very expensive. The microbial growth in solid-state fermenters is determined by the intake of oxygen and the evolution of carbon dioxide (Pandey et al. 2007).

11.3 Recovery and Purification of Beta-Glucanases

The beta-glucanases are produced as extracellular enzymes (secreted into the culture medium) or as intracellular enzymes (present within the cells). Crude or highly purified commercial enzymes in solid or liquid forms are available based on necessity. Recovery and purification have different steps (Fig. 11.2). It is mainly based on the enzyme's nature and desired purity and to minimize the loss of activity. Recovery of the extracellular enzyme is easier than intracellular or cell-bound enzymes. For the recovery of intracellular enzymes cell disruption through enzymatic treatments (lysozyme), sonication, glass beads or using high pressure is required. Purification cost depends on the further application of enzymes. It is very high and covers more than 50% of the total production cost. The first step in purification involves filtration or centrifugation for removing the cell debris. Ultrafiltration is used for concentrating the enzyme solution. Removal of impurities such as other proteins, polysaccharides and nucleic acid is necessarily increasing the activity and is important for its use in animal feed and breweries. Chromatographic techniques like size exclusion, hydrophobic interaction ion-exchange and affinity chromatography are generally used for the purification of enzymes. Depending on the enzyme's surface charge, molecular size and hydrophobicity different chromatographic adsorbents are used for separation (Singhania et al. 2010). After purification and concentration, the final product is obtained in dried powder form. This can be achieved by lyophilizers or film evaporators.

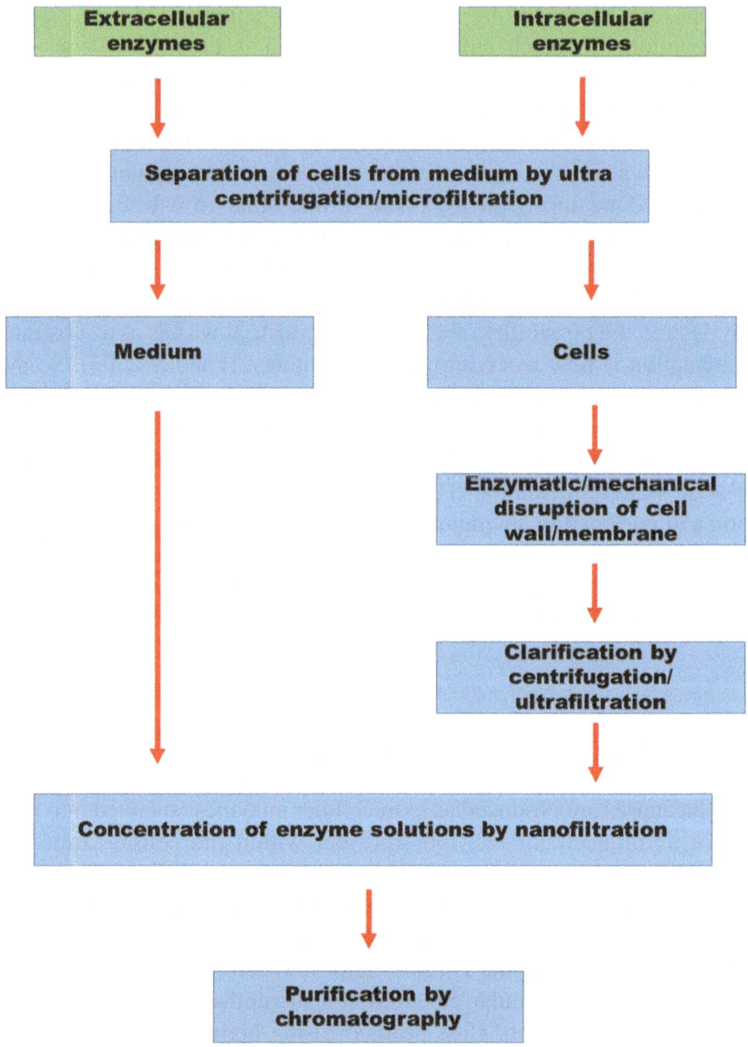

Fig. 11.2 Steps involved in the recovery of enzymes

11.4 Production of Recombinant Beta-Glucanases

Usually, for increasing productivity or enzymatic properties, genetic engineering techniques are used. Altering the host organism's genomic region that may result in the enzyme's downregulation of expression or deletion/destabilization, as well as the precise organization of the gene that codes for the enzyme, in order to improve the stability, activity, and yield of the resulting protein. Recombinant enzymes are more economical and cost-effective because of the ease of purification (Jaeger et al. 2001).

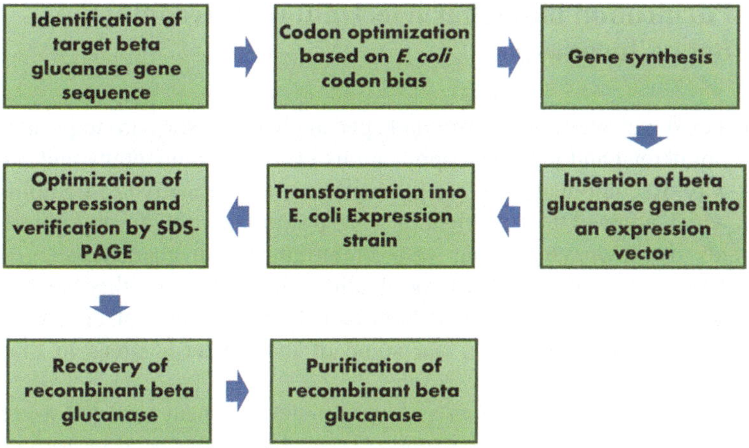

Fig. 11.3 Steps involved in the recombinant production of beta-glucanases

At present, more than 90% of commercially available high-value enzymes are recombinant enzymes. A widely used organism for the recombinant expression of beta-glucanases is *E. coli* (Palumbo et al. 2003; Chen et al. 2019; Nagata et al. 1990; Borriss et al. 1985). *E. coli* can easily be grown in an inexpensive culture media. The genetic manipulation in *E. coli* is easy and the available strains have fewer proteolytic activities. It produces high levels of protein as it grows with high cell densities. However, *E. coli* does not secrete the enzyme into the culture media. Sometimes, protein may be inactive due to aggregation and forms inclusion bodies. Truncated proteins may produce as a result of codon bias. The strains are unable to do some post-translational modifications of eukaryotic proteins. Though, these disadvantages do not preclude the recombinant expression of proteins in the *E. coli* host. Around one-third of FDA approved recombinant therapeutic proteins are synthesized in *E. coli*. The overall process of recombinant production of beta-glucanase is depicted in Fig. 11.3.

An extensively employed affinity purification technique is the IMAC (Immobilized Metal Ion Chromatography), which exploits the interaction of the amino acid Histidine with different metal ions like Ni^{2+}, Co^{2+}, Zn^{2+} and Cu^{2+}. The most widely used combination is a histidine tag with the metal Ni^{2+}. This allows the purification of recombinant proteins containing short polyhistidine fusion tags (Bornhorst and Falke 2000). The purifications can be accomplished as a single step in one day time with a significant improvement above the extended purification protocol of the non-histidine-tagged protein (Trigoso et al. 2016). Usually, the affinity tags composed of six histidine residues are long enough to yield better interactions with the affinity matrix, ensuring 100-fold enrichments in a single purification step and the purity can be achieved up to 95% (Janknecht et al. 1991). It has been successfully employed in a variety of expression systems such as *E. coli* (Van Dyke et al. 1992), *Saccharomyces cerevisiae* (Li et al. 2012), insect cells (Kollewe and Vilcinskas 2013) and mammalian cells (Cass et al. 2005).

11.5 Purification and Characterization of Recombinant Beta-Glucanases

Microbial enzymes with high purity and special characteristics are important to use in their commercial and industrial applications in an environmentally and economic friendly way. Recombinant enzyme production is an alternative to traditional production, which allowed the modification of a specific microorganism to produce the high yields of an enzyme along with preferred special characteristics such as thermotolerance, tolerance to acidic or alkaline conditions and other harsh reaction conditions in industries. The purification and the exploration of physio-chemical characteristics of novel enzymes are some of the advanced areas in bioprocess research.

The purification of *Streptomyces* origin recombinant exo-β-1,4-glucanase (exo14) and endo-β-1,3-glucanase (endo13) enzymes have been performed using immobilized metal-affinity chromatography. The histidine-tagged recombinant proteins expressed in BL21 Star™ (DE3) One Shot® *E. coli* cells were purified using simple, effective and fast Ni-NTA affinity chromatography. Both proteins were eluted using an elution buffer containing 250 mM imidazole. The expressed recombinant proteins with 6-His tag are precisely bound by chelated divalent nickel ions, and eluted by the addition of imidazole through the competition or protonation of histidine residues by the reduction in pH (Smith et al. 1988). The single IMAC purification step produced the most efficient and almost complete separation of proteins with fewer contaminants. The purified exo14 and endo13 protein fractions showed homogenous protein bands with a respective molecular mass of 70 kDa and 44 kDa (Fig. 11.4), which correspondingly yielded the total activities as 57% and 67% when considering the total activities in crude soluble fraction as 100%. By correlating the results with the previous studies, the purified His-tagged recombinant fusion proteins from *Streptomyces* sp. yielded the highest activities than native proteins (Nishimura et al. 2001). HiTrap SP and Q column purified native endo-β-1,3-glucanase protein from *Streptomyces sioyaensis* yielded only 15–25% of total activity (Hong et al. 2002). The highest reported activities from *Streptomyces* spp. were achieved for exo14 and endo13, which were 72 U mg-1 and 65.63 U mg-1, respectively. The details of purification are summarized in Table 11.1.

Recombinant exo14 and endo13 exhibited remarkable biochemical characteristics. The optimum activity of exo14 was recorded at a temperature 50 °C which also maintains more than 60% of activity at 30–60 °C. Exo-β-1,4-cellobiohydrolase from *Streptomyces coelicolor* A3(2) showed maximum activity towards Avicel at 50 °C (Lee et al. 2018). The endo13 recorded optimum activity at 60 °C, also maintained more than 80% of activity at 30 to 60 °C. Endo-β-1,3-glucanase from *Streptomyces* sp. Mo also showed the optimum activity towards Curdlan at 60 °C (Kurakake et al. 2013). The residual activity of exo14 at 50 and 60 °C showed more than 60%, whereas the endo13 showed more than 70% activities at 50 and 60 °C after 6 h, retained 30% of enzyme activity at 70 °C after 5 h and also maintained more than 10% of its activity 80 °C after 4 h of incubation (Edison et al. 2020). According to

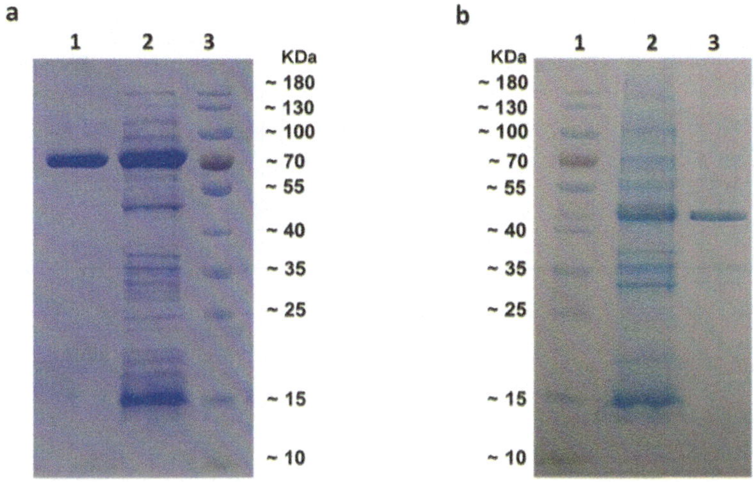

Fig. 11.4 SDS-PAGE analysis of purified recombinant proteins. (**a**) SDS-PAGE analysis of purified exo-β-1,4-glucanase (exo14) protein. Lane 1: Nickel affinity-purified exo14 protein, Lane 2: Soluble total protein of cell lysate after induction, Lane 3: Protein molecular weight marker. (**b**) SDS-PAGE analysis of purified endo-β-1,3-glucanase (endo13) protein. Lane 1: Protein molecular weight marker, Lane 2: Soluble total protein after induction, Lane 3: Nickel affinity-purified endo13 protein

Table 11.1 Purification of exo-β-1,4-glucanase (exo14) and endo-β-1,3-glucanase (endo13) proteins

optexo14						
Step	Volume (mL)	Total activity (U.mL^{-1})	Total protein (mg.mL^{-1})	Specific activity (U. mg^{-1})	Yield (%)	Purification fold
Crude extract	5	5300	830	6.3	100	1
Ni^{2+} affinity column	1	3024	42	**72.0**	57.0	11.42
optendo13						
Crude extract	5	1076	322	3.34	100	1
Ni^{2+} affinity column	1	722	11	**65.63**	67.1	19.64

Cano-Ramírez et al. (2016), bacteria β-glucanases showed activity at optimal pH, but do not show activity at wide pH intervals. Apart from these reports *Streptomyces* spp. established an exception that the enzymes exo14 and endo13 showed significant activity and stability in wide ranges of pH from 4.0 to 8.0. These findings were comparable to the exo-β-1,4-glucanase from *Trichoderma viridi* (Irshad et al. 2012) and *Chaetomium olivaceum* (El-Gindy et al. 2003) and endo-β-1,3-glucanase from *Trichoderma harzianum* (Noronha and Ulhoa 1996) and *Trichoderma koningi*

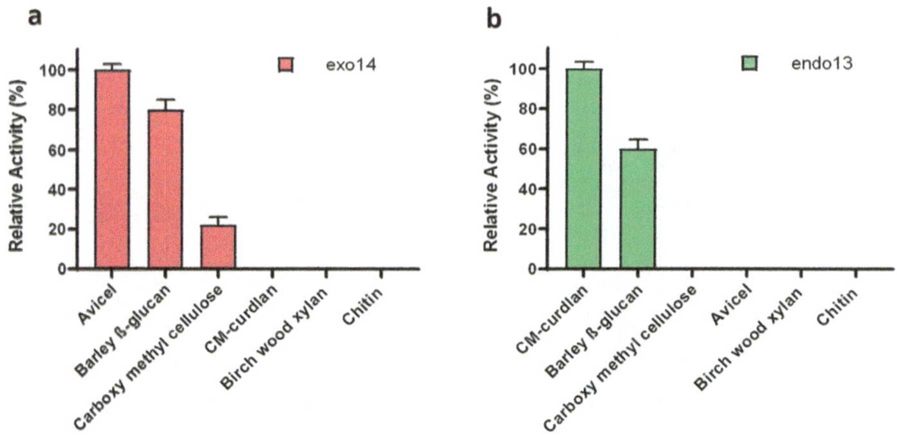

Fig. 11.5 Analysis of substrate specificity of recombinant beta-glucanase enzymes (**a**) exo14 and (**b**) endo13 against various substrates. The activity was assayed using Avicel, barley β-glucan, carboxymethylcellulose, CM-Curdlan, birchwood xylan and chitin as substrates, at a final concentration of 0.5% under optimum conditions. The values were expressed as relative activities (%) to the maximal activity. The maximum activity was defined as 100%

(Monteiro and Ulhoa 2006) confirmed exo14 and endo13 from *Streptomyces* spp. have significant more thermostability and pH stability than fungal enzymes.

The substrate specificity studies of recombinant exo14 showed high affinity towards cellulolytic substrate Avicel, barley β-glucan and carboxymethyl cellulose. Endo13 showed high specificity towards the β-1,3 linked substrate CM-Curdlan, and also showed 60% specificity towards barley β-glucan which confirmed the enzyme is also able to cleave the internal β-1,3-linkages within the barley β-glucan chain, defines the enzyme is endo-β-1,3-glucanase (Fig. 11.5). As per the previous reports, the high affinity of enzymes towards Avicel confirms it as an exo-acting enzyme and the Avicel is used as a model substrate for detecting the cellobiohydrolase (CBH) activity (Kruus et al. 1995; Teeri 1997; Percival Zhang et al. 2006). The specificity towards insoluble substrates like Avicel and barley β-glucan is enabled by the carbohydrate-binding module of β-glucanase, which primarily recognize the glucan chain and resulting effective hydrolysis of substrates (Boraston et al. 2004).

It is commonly accepted that some metal ions and inhibitors significantly alter enzyme activities (Wang et al. 2012). The metal ions such as Cu^{2+}, Co^{2+} and Fe^{2+} showed significant inhibitory effects on exo14 and endo13. Zn^{2+} produced a highly detrimental effect on exo14 activity (Fig. 11.6. a). The inhibitory action of Zn^{2+} and Cu^{2+} is a general feature of β-glucanases (Cano-Ramírez et al. 2016). According to Lim et al. (2016), Mn^{2+} and Cu^{2+} were reported as fungal β-glucanase inhibitors. However, Mn^{2+} stimulated the activities of both exo14 and endo13 *Streptomyces* enzymes. Inhibition by divalent metal ions has been reported widely in many glycosyl hydrolases, might be by attacking the thiol groups at the active site, however, this necessitates additional validation. On the other hand, both enzyme activities are inhibited by EDTA, DTT, β-mercaptoethanol and SDS (Fig. 11.6).

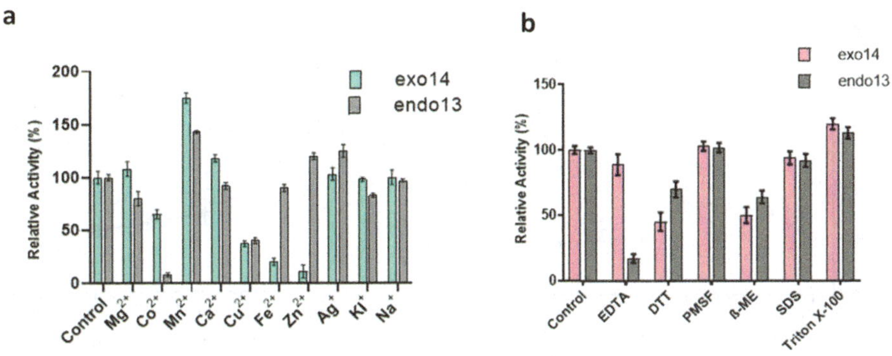

Fig. 11.6 (**a**) Effect of metal ions on the recombinant exo14 and endo13 activities. The enzymes were pre-incubated with 5 mM of $MgSO_4$, $CoCl_2$, $MnCl_2$, $CaCl_2$, $CuSO_4$, $FeSO_4$, $ZnSO_4$, $AgNO_3$, KI and NaCl. The enzyme activities were determined by standard assay under optimum conditions. (**b**) Effect of inhibitors on the recombinant exo14 and endo13 activities. The enzymes were pre-incubated with 2 mM of EDTA, DTT, PMSF, β-ME and SDS. The enzyme activities were determined by standard assay under optimum conditions. The values were expressed as relative activities (%) to the activity of control defined as 100%

These results indicate that metal ions and disulphide linkages are necessitated for the catalytic activities (Wang et al. 2014). As shown in our results, Triton X-100 stimulated the β-1,3-1,4-glucanase activity in *Bacillus amyloliquefaciens* (Sun et al. 2012).

According to the results of kinetic studies, the purified recombinant exo14 and oendo13 showed Km values of 7.508 and 9.221 mM using the substrate barley β-glucan. The lower Km values designate the higher affinity to β-glucan substrates which proves the beneficial application of both enzymes in industrial sectors such as brewing and animal feed enzyme industries (Edison et al. 2020). Similar observations were reported by Wang et al. (2014), in β-1,3-1,4-glucanase from *Bacillus tequilensis*. The compounds such as ethanol, sulphite, ferulic acid and gallic acid are produced in breweries naturally as a result of fermentation. The influence of these compounds on exo14 and endo13 activities revealed that the individual enzymes, as well as the cocktail mix of both enzymes, were active at different concentrations of ethanol, ferulic acid and gallic acid. However, a 30% of loss in the relative activity of individual enzymes as well as less than 5% of loss in the activity of cocktail mix were detected (Fig. 11.7). The β-1,3-Glucanase from *Delftia tsuruhatensis* Strain MV01 also showed similar activities in the above-mentioned conditions (Blättel et al. 2011).

The purification of recombinant exo14 and endo13 enzymes using Ni^{2+} charged IMAC and followed by the study of the key physio-chemical characteristics such as the optimum pH and temperature conditions, stability under varying pH and temperatures, specificity to different substrates, the effect of metal ions and inhibitors and kinetic parameters such as Vmax and Km confirmed the efficient use of enzymes in industrial applications. The influence of selected brewery-relevant parameters such as ethanol, sulphites, ferulic acid and gallic acid on enzyme activities

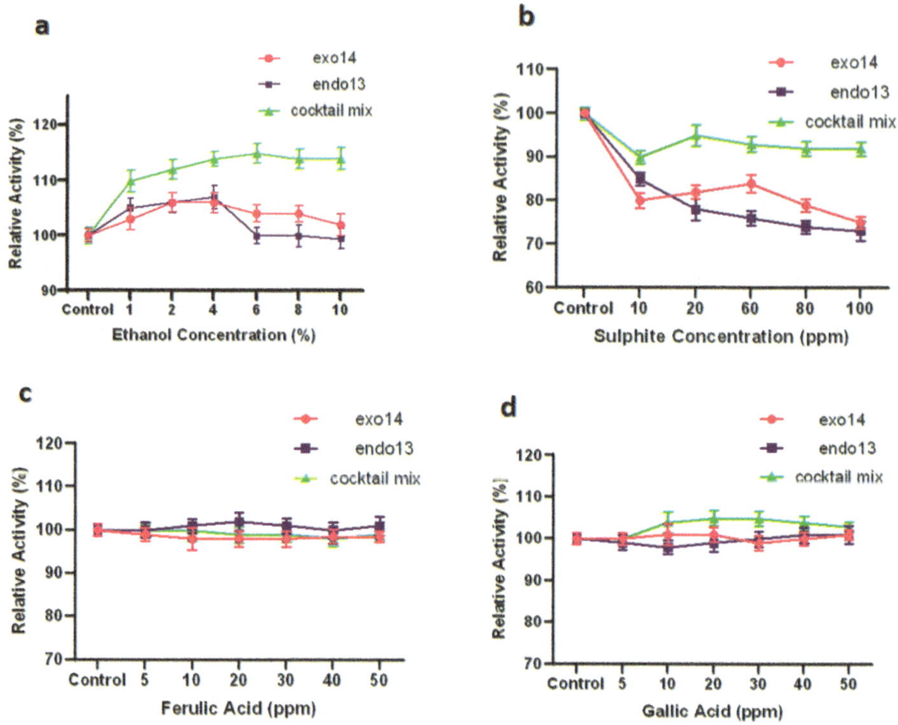

Fig. 11.7 Influence of brewery relevant parameters on individual and cocktail mix of exo14 and endo13 activities on varying concentrations of (**a**) Ethanol (1–10%), (**b**) Sulphite (10–100 ppm), (**c**) Ferulic acid (5–50 ppm) and (**d**) Gallic acid (5–50 ppm). The values were expressed as relative activities (%) to the activity of control defined as 100%

recommended the enzymes can be used as better candidates in the brewing industry as well as feed enzyme industry (Edison et al. 2020).

11.6 Conclusions

Enzyme industry which supports many industries such as food, health, paper and cloth has a huge market share (52.1% of the total revenue in 2019) and is predicted to increase with a compound annual growth rate of 6.9%. The research sector in the enzyme industry is always looking for more promising molecules/approaches to increase the efficiency of the enzyme in terms of activity, shelf life and production cost. Thus, the search for new strains and genes is always "ON" with those enzymes which have a wider application in many industries. Being an enzyme with multiple applications beta-glucanase enzyme was identified and characterized by a number of microbial sources. Many commercially available beta-glucanases are produced by the strains identified from nature along with the heterologous expression using the lab strains. Some of the factors which affect the synthesis, stability and purification, when it is expressed using a natural strain can be avoided when a heterologous expression system is used. The engineered lab strains for protein expression reduce the negative effects to the maximum so that the protein expressed will be stable and active and will have a high level of expression. *E. coli* being one of the favourable hosts for protein expression was used as an expression system and two proteins with beta-glucanase activity were expressed, and purified. The improvements in activity and stability can be achieved further by careful protein engineering in which the residues which interfere with the activity/stability can be modified without affecting the overall structure/function of the protein. Using the same gene in some other expression host also will be a feasible approach to ponder off.

References

Aono R, Hammura M, Yamamoto M, Asano T (1995) Isolation of extracellular 28 and 42-Rilodalton β-1,3- glucanases and comparison of β-1,3-glucanase produced by Bacillus circulans I IAM1165. Appl Environ Microbiol 61(1):122–129

Bajomo MF, Young TW (1992) Development of mashing profile for the use of microbial enzymes in brewing with raw sorghum (80%) and malted barley or sorghum malt (20%). J Inst Brew 98:516

Blättel V, Larisika M, Pfeiffer P, Nowak C, Eich A, Eckelt J, König H (2011) Beta-1,3-glucanase from Delftia tsuruhatensis strain MV01 and its potential application in vinification. Appl Environ Microbiol 77(3):983–990

Boraston AB, Bolam DN, Gilbert HJ, Davies GJ (2004) Carbohydrate-binding modules: fine tuning polysaccharide recognition. Biochem J 382(3):769–781

Bornhorst JA, Falke JJ (2000) Purification of proteins using polyhistidine affinity tags. Methods Enzymol 326:245–254

Borriss R, Bäumlein H, Hofemeister J (1985) Expression in Escherichia coli of a cloned β-glucanase gene from Bacillus amyloliquefaciens. Appl Microbiol Biotechnol 22:63–71

Burton SG, Cowan DA, Woodley JM (2002) The search for the ideal biocatalyst. Nat Biotechnol 20(1):37–45

Cano-Ramírez C, Santiago-Hernández A, Rivera-Orduña FN, García-Huante Y, Zúñiga G, Hidalgo-Lara ME (2016) Expression, purification and characterization of an endoglucanase from Serratia proteamaculans CDBB-1961, isolated from the gut of Dendroctonus adjunctus (Coleoptera: Scolytinae). AMB Express 6(1):63

Cass B, Pham PL, Kamen A, Durocher Y (2005) Purification of recombinant proteins from mammalian cell culture using a generic double-affinity chromatography scheme. Protein Expr Purif 40(1):77–85

Chen F, Ye J, Sista Kameshwar AK, Wu X, Ren J, Qin W, Li DW (2019) A novel cold-adaptive endo-1,4-β-Glucanase from Burkholderia pyrrocinia JK-SH007: gene expression and characterization of the enzyme and mode of action. Front Microbiol 10:3137

Edison LK, Dan VM, Reji SR, Pradeep NS (2020) A strategic production improvement of Streptomyces Beta glucanase enzymes with aid of codon optimization and heterologous expression. Biosci Biotechnol Res Asia 17(3):587–599

El-Gindy AA, Saad RR, Fawzi E (2003) Purification and some properties of exo-1,4-beta-glucanase from Chaetomium olivaceum. Acta Microbiol Pol 52(1):35–44

Farinas ET, Bulter T, Arnold FH (2001) Directed enzyme evolution. Curr Opin Biotechnol 12(6): 545–551

Hölker U, Höfer M, Lenz J (2004) Biotechnological advantages of laboratory-scale solid-state fermentation with fungi. Appl Microbiol Biotechnol 64(2):175–186. https://doi.org/10.1007/s00253-003-1504-3. Epub 2004 Feb 13

Hong TY, Cheng CW, Huang JW, Meng M (2002) Isolation and biochemical characterization of an endo-1,3-β-glucanase from Streptomyces sioyaensis containing a C-terminal family 6 carbohydrate-binding module that binds to 1,3-β-glucan. Microbiology (Reading) 148(4):1151–1159

Irshad M, Anwar Z, Afroz A (2012) Characterization of exo−1,4-β glucanase produced from Trichoderma Viridi through solid-state bio-processing of orange peel waste. Adv Biosci Biotechnol 3:580–584

Jaeger KE, Eggert T, Eipper A, Reetz MT (2001) Directed evolution and the creation of enantioselective biocatalysts. Appl Microbiol Biotechnol 55(5):519–530

Janknecht R, de Martynoff G, Lou J, Hipskind RA, Nordheim A, Stunnenberg HG (1991) Rapid and efficient purification of native histidine-tagged protein expressed by recombinant vaccinia virus. Proc Natl Acad Sci U S A 88(20):8972–8976

Kollewe C, Vilcinskas A (2013) Production of recombinant proteins in insect cells. Am J Biochem Biotechnol 9(3):255–271

Kruus K, Wang WK, Ching JT, Wu JHD (1995) Exoglucanase activities of the recombinant clostridium thermocellum CelS, a major cellulosome component. J Bacteriol 177(6):1641–1644

Kurakake M, Yamanouchi Y, Kinohara K, Moriyama S (2013) Enzymatic properties of β-1,3-glucanase from Streptomyces sp Mo. J Food Sci 78(4):C502–C506

Lee CR, Chi WJ, Lim JH, Dhakshnamoorthy V, Hong SK (2018) Expression and characterization of the processive exo-β-1,4-cellobiohydrolase SCO6546 from Streptomyces coelicolor a(3). J Basic Microbiol 58(4):310–321

Li S, Clemençon B, Catty P, Brandolin G, Schlattner U, Rousseau D (2012) Yeast-based production and purification of HIS-tagged human ATAD3A, a specific target of S100B. Protein Expr Purif 83(2):211–216

Lim JH, Lee CR, Dhakshnamoorthy V, Park JS, Hong SK (2016) Molecular characterization of Streptomyces coelicolor a (3). FEMS Microbiol Lett 363(3):245

Monteiro VN, Ulhoa CJ (2006) Biochemical characterization of a beta-1,3-glucanasefrom Trichoderma koningii induced by cell wall of Rhizoctonia solani. Curr Microbiol 52(2):92–96

Nagata S, Maru I, Ishihara F, Misono H, Nagasaki S (1990) Cloning and expression of endo-beta-1,3-glucanase gene from Flavobacterium dormitator in Escherichia coli and characterization of the gene product. Agric Biol Chem 54(10):2675–2680

Nishimura T, Bignon C, Allouch J, Czjzek M, Darbon H, Watanabe T, Henrissat B (2001) Streptomyces matensis laminaripentose hydrolase is an 'inverting' beta-1,3-glucanase. FEBS Lett 499(1–2):187–190

Noronha EF, Ulhoa CJ (1996) Purification and characterization of an endo-β-1,3-glucanase from Trichoderma harzianum. Can J Microbiol 42(10):1039–1044

Oshima K, Itaya S (1926) A new method for quantitative estimation of starch by Aspergillus amylase (taka-diastase). J Agric Chem Soc Jpn 2(9):122–123

Palumbo JD, Sullivan RF, Kobayashi DY (2003) Molecular characterization and expression in Escherichia coli of three beta-1,3-glucanase genes from Lysobacter enzymogenes strain N4-7. J Bacteriol 185(15):4362–4370

Pandey A, Soccol CR, Larroche C (2007) Introduction. In: Pandey A, Soccol CR, Larroche C (eds) Current developments in solid-state fermentation. Springer Science, New Delhi.: New York and Asiatech Publishers, Inc, pp 3–12

Pandey A (2003) Solid-state fermentation. Biochem Eng J 13:81–84

Percival Zhang YHP, Himmel ME, Mielenz JR (2006) Outlook for cellulase improvement: screening and selection strategies. Biotechnol Adv 24(5):452–481

Powell KA, Ramer SW, Del Cardayré SB, Stemmer WPC, Tobin MB, Longchamp PF, Huisman GW (2001) Directed evolution and biocatalysis. Angew Chem Int Ed Engl 40(21):3948–3959

Shao-jun D, Wei G, Buswell JA (2001) Endoglucanase 1 from the edible straw mushroom *Volvariela volvacea*. Purification, characterization, cloning and expression. Eur J Biochem 268:5687–5695

Scopes RK (1995) Overview of protein purification and characterization. Current protocols in. Protein Sci 00:1.1.1–1.1.6

Singhania RR, Patel AK, Pandey A (2010) The industrial production of enzymes. Ind Biotechnol:207–225

Smith MC, Furman TC, Ingolia TD, Pidgeon C (1988) Chelating peptide immobilized metal ion affinity chromatography. A new concept in affinity chromatography for recombinant proteins. J Biol Chem 263(15):7211–7215

Srinivasa M (2005) Studies on the extraction and characterization of thermostable a-amylase from pericarp of Borassus indica. Afr J Biotechnol 4(3):289–291

Stahmann KP, Pielken P, Schimz KL, Sahm H (1992) Degradation of extracellular β-(1,3), (6)-O-Dglucanase by Botrytis cinera. Appl Environ Microbiol 58(10):3347–3354

Sun JT, Wang HX, Lu WP, Ma CY, Lou ZX, Yao H, Dai Y (2012) Cloning and expression of a thermostable beta-1,3-1,4-glucanase from Bacillus amyloliquefaciens ATCC 23350. Ann Microbiol 62:1235–1242

Tang-Yao H (2002) Isolation and biochemical characterization of an endo-1, 3- β-glucanase from Streptomyces sioyaensis containing a C-terminal family 6 carbohydrate binding 1,3- β-glucan. Microbiol 148:1151–1159

Teeri TT (1997) Crystalline cellulose degradation: new insight into the function of cellobiohydrolases. Trends Biotechnol 15(5):160–167

Trigoso YD, Evans RC, Karsten WE, Chooback L (2016) Cloning, expression, and purification of histidine-tagged Escherichia coli dihydrodipicolinate reductase. PLoS One 11(1):e0146525

Van Dyke MW, Sirito M, Sawadogo M (1992) Single-step purification of bacterially expressed polypeptides containing an oligo-histidine domain. Gene 111(1):99–104

Wang G, Xiaowen Z, Li W, Keke W, Fanglin P, Linsong W (2012) The activity and kinetic properties of cellulases in substrates containing metal ions and acid radicals. Adv Biol Chem 2: 390–395

Wang J, Niu C, Liu X, Chen X, Li Q (2014) Characterization of a new 1,3-1,4-β-glucanase gene from Bacillus tequilensis CGX5-1. Appl Biochem Biotechnol 173(3):826–837

Chapter 12
Biotechnological Avenues in Hybrid Beta-Glucanases

Lekshmi K. Edison, K. Anu, and N. S. Pradeep

Abstract Cereal beta-glucans are composed of multiple glycosidic linkages. Hence, the complete hydrolysis of beta-glucan substrates in breweries and animal feed enzyme industries requires the interaction of different beta-glucanase enzymes. Recent advances in creating hybrid enzymes pointed towards the development of hybrid beta-glucanases. Hybrid enzyme technology is an important aspect of protein engineering and covers a broad range of significant and potential applications which will expand the use of enzymes in biotechnological industries. Hybrid beta-glucanases are composing the fusion polypeptides, which are designed by the physical connection of enzymatic domain segments. This approach provided significant success because individual domains do not normally interact with the rest of the enzymatic part. The detailed knowledge of available structure–function relationships of enzymes and innumerable genetic engineering techniques help to develop more sophisticated designs of industrially important hybrid beta-glucanases. It involves the designing of combined multienzyme complexes and also manipulating the regulation of enzymatic activities. This chapter explains the role of protein engineering in hybrid enzyme production and the multifunctionality of hybrid beta-glucanases.

Keywords DNA shuffling · Seamless cloning · Retrofitting · Exo14-endo13 hybrid beta-glucanase

---

L. K. Edison
Rajiv Gandhi Centre for Biotechnology (RGCB), Thiruvananthapuram, Kerala, India

K. Anu
Chinmaya Arts and Science College for Women, Kannur, Kerala, India

N. S. Pradeep (✉)
KSCSTE-Malabar Botanical Garden and Institute for Plant Sciences, Kozhikode, Kerala, India

N. S. Pradeep, L. K. Edison (eds.), *Microbial Beta Glucanases*, Interdisciplinary Biotechnological Advances, https://doi.org/10.1007/978-981-19-6466-4_12

12.1 Introduction

One of the major avenues in biotechnology and protein biochemistry is the production of hybrid enzymes, by joining the structural elements of different enzymes. The engineering enzymes have novel or modified properties such as catalytic property and substrate-binding affinity and are one of the most exhilarating developments in protein engineering. Capabilities of computer modelling using some bioinformatics tools allow the designing of hybrid enzymes. Recombinant DNA technology along with in vitro DNA synthesis makes the hybrid or chimeric design of enzymes possible. Hybrid biocatalysts are designed by selecting DNA sequences from more than one parent, then manipulating the DNA fragments to generate constructs encoding desired catalytic and physical properties. The strategies for the production of hybrid enzymes involve the creation of new enzymes and the modification of pre-existing enzymes. The majority of hybrid enzymes are the modified enzymes of pre-existing ones because the recombination of pre-existing segments of enzymes generates enzymes with novel properties (Béguin 1999). Protein engineering permits the creation of hybrid enzymes through the introduction of individual amino acid residues, whole subdomains, partial subdomains, secondary structural elements or whole proteins. Some enzymes are generated by exchanging the shorter elements between homologous proteins. It retains the original function of enzymes but modifies the properties such as thermostability, pH stability and optimum kinetic parameters and substrate specificity. Another criterion is the joining of two or more functional units of different enzymes to create bi- or multifunctional proteins (Nixon et al. 1998). Hybrid enzymes possess some improved or novel properties.

The major reasons for creating hybrid proteins are due to the poor stability and low catalytic efficiency of existing proteins under industrial conditions, such as in combination with organic solvents or at high temperatures or pH in fermenters. In some other cases, heterologous host exhibits ineffective folding and susceptibility to proteases causing defective catalytic functions and truncated versions of proteins (Corey and Corey 1996). Currently available structural knowledge and detailed sequence information guide to combining the recognized functions and properties of existing enzymes and generates chimeric or hybrid engineered enzymes. It alters the enzymatic and non-enzymatic properties of natural enzymes, further expanding the potential use of the enzyme. The exchanges and combining of multiple domains of different enzymes generally resulted in properties between the parent enzymes. The hybrids of beta-glucosidase from *Agrobacterium tumefaciens* (pH 7.2–7.4 and temperature 60 °C) and *Cellvibrio gilvus* (pH 6.2–6.4 and 35 °C) showed activity at pH 6.6–7.0 and temperature 45–50 °C. Km values also showed in-between the parent enzymes (Singh and Hayashi 1995). The creation of hybrid enzymes also enhances the production and folding. Researchers use the term 'retrofitting' for the process of adding new elements to an existing enzyme to alter its functionality through the exchange of domains, introducing secondary structural elements or by introducing potential amino acids or mutations (Hecht 1996). The major protein engineering goal to create the hybrid enzymes are modulating the specificity or catalytic function of existing enzymes, introducing catalytic amino acids to improve the catalysis and introducing binding and folding properties (Hedstrom et al. 1992).

12.2 Role of Protein Engineering in Hybrid Enzyme Production

The recombinant DNA technology successively decreases the cost of industrial production of enzymes. However, most of the enzymes are not stable in harsh reaction environments such as acidic pH, high temperature and presence of organic solvents. Protein engineering allows to change the amino acid sequence of the enzyme to achieve the results like specificity to a broad range of substrates, increases temperature and pH stability and stability with different organic solvents. Protein engineering has a vital role in basic research, therapeutic research, biotechnology, material sciences and so on. In protein engineering, the application of different genomic and post-genomic strategies along with in vitro *synthesis* increase the functional properties of enzymes which carry unique physical, chemical and biological properties (Kohrer and Rajbhandary 2009; Turanli-Yildiz et al. 2012). Immense knowledge of amino acid sequences, three-dimensional structures and mechanism of enzymatic action improves the alteration or modification of enzymes through protein engineering.

Protein engineering methods are grouped into two categories—rational design and combinatorial methods. Rational design created a small number of enzyme variants through amino acid substitutions and site-directed mutagenesis. It requires only a little effort to improve the enzyme qualities. It helps to study the structure–function relationship and improves the enzyme properties like thermoactivity, foldability, specificity, and selectivity. On the other hand, the combinatorial method including circular permutation, homologous and nonhomologous recombination or random insertion produces a large number of enzyme variants (Singhania et al. 2010). Hybrid enzymes are synthesized through domain swapping or by the end-to-end ligation of biological or chemically synthesized whole or partial domains (Meunier 2008; Palmer et al. 2011; Kim and Izumi 2014; Balabanova et al. 2014). Existing computational tools of structure prediction and other structural principles implemented from natural proteins permit the prediction or introduction of new structural modifications of enzymes (Basit and Wechsler 2011). Homology modelling and docking of enzymes with substrate or inhibitor molecules can simplify the prediction of hybrid enzymes. In vitro synthesis of predicted enzymes can be realized through recombinant production in a heterologous host (Pronk et al. 2013; Koga et al. 2012; Amin et al. 2013; Golotin et al. 2015).

12.3 Methods for the Creation of Hybrid Enzymes

Hybrid enzymes are a result of precise and defined protein engineering. The seamless joining of sequences by routine gene cloning can be used to combine two or more, whole or partial domains of enzymes. However, it may be impossible sometimes, when there is altered or extra amino acids at the junction of two sequences (Sinha et al. 1996) though, many techniques have been developed to avoid this. Seamless fusion can be created by splicing and overlap extension through polymerase chain reaction (Warrens et al. 1997). Moreover, cloning vectors can construct with two or more fused sequences created by unidirectional deletions of intervening bases between them (Kim et al. 1991). Homologous recombination is another method to generate nonspecific hybrids (Schneider et al. 1981). Some other hybrid enzymes have been created by the random approach of DNA shuffling, which is a directed evolution method. The gene is randomly fragmented and reassembled by error-prone PCR. It is used to shuffle homologous genes for creating a hybrid enzyme library (Stemmer 1994). Hybrid enzymes are categorized into three different types based on their mode of action. In the first type, it points to a single target. But the different domains are active at different times. All the domains of the hybrid enzymes interact with the same target. In the second category, different domains of the hybrid enzymes act independently on different and nonrelated targets. Finally, category three acts on two different and related targets on the same molecule. Both domains act at the same time. However, currently, the hybrid enzymes with a different mode of action have more interest in industrial applications and also in drug discovery (Meunier 2008).

12.4 Beta-Glucanase Hybrid Enzymes: Multifunctionality

The hybrid enzyme is a chimeric multifunctional enzyme created from two or more unrelated DNA sequences. Hybrid enzymes are synthesized by the ligation of functional domain fusion biological or chemically synthesized DNA fragments (Meunier 2008; Balabanova et al. 2014). Hybrid enzymes with multiple domains have different functional activities or modes of action (Reddy et al. 2009; Furtado et al. 2013; Rizk et al. 2015; Otsuka et al. 2015). Functional domains of hybrid enzymes are covalently linked through a short amino acid sequence called linkers or spacers (Reddy Chichili et al. 2013). Two types of linkers used for making hybrid enzymes are rigid linkers and flexible linkers. Rigid linkers are proline-rich and prevent the undesired interactions between distinct domains. Flexible linkers are glycine-rich and it connects the functional domains without disturbing the functions of each domain. Linkers enhance the solubility, functionality and foldability of hybrid enzymes (Balabanova et al. 2014; Reddy et al. 2009; Reddy Chichili et al. 2013).

Beta-glucanases are the glucanolytic enzymes, which have major roles in beer production and animal feed production. These enzymes are used in the efficient hydrolysis of beta-1,3-1,4-linked cereal beta-glucans in the brewing industry and feed enzyme industry. Hence, the complete hydrolysis of cereal beta-glucan requires the synergic action of different beta-glucanases. Edison and Pradeep., 2020, reported isolation of exo-β-1,4-glucanase (exo14) and endo-β-1,3-glucanase (endo13) producing prominent *Streptomyces* strains, respectively *Streptomyces althioticus* TBG-MR17 and *Streptomyces cinereoruber* sub sp. *cinereoruber* TBG-AL13 from the Western Ghats regions of Kerala. It was the first report of the exploration of Western Ghats actinomycetes for both exo-β-1,4-glucanase and endo-β-1,3-glucanase enzymes with very high activities. The isolated enzymes' good characteristic has suggested both the enzymes individually as well as cocktail mix, will be the better candidates for applications in the brewing industry and animal feed enzyme industry for the efficient hydrolysis of cereal β-glucan (Edison et al. 2020a). For the functional characterization, the structural domain architecture of both enzymes has been predicted through in silico analysis (Edison et al. 2020b). Based on the structural perspectives, an 'exo14-endo13' hybrid enzyme structure was predicted through in silico end-to-end fusion of catalytic domains of exo-β-1,4-glucanase (exo14), a catalytic domain and carbohydrate-binding module (CBM) of endo-β-1,3-glucanase (endo13), using Chimera Version 1.12. For extracellular secretion, signal peptide sequences from exo-β-1,4-glucanase (exo14) were maintained at the N-terminal end of the hybrid enzyme. Both catalytic domains and CBM were separated using native linker sequences from the enzymes themselves. The predicted exo14-endo13 hybrid enzyme structure is depicted in Fig. 12.1 (unpublished data). The substrate-binding affinity of the hybrid enzyme was confirmed by in silico docking with substrates such as Cellotriose (for exo14) and

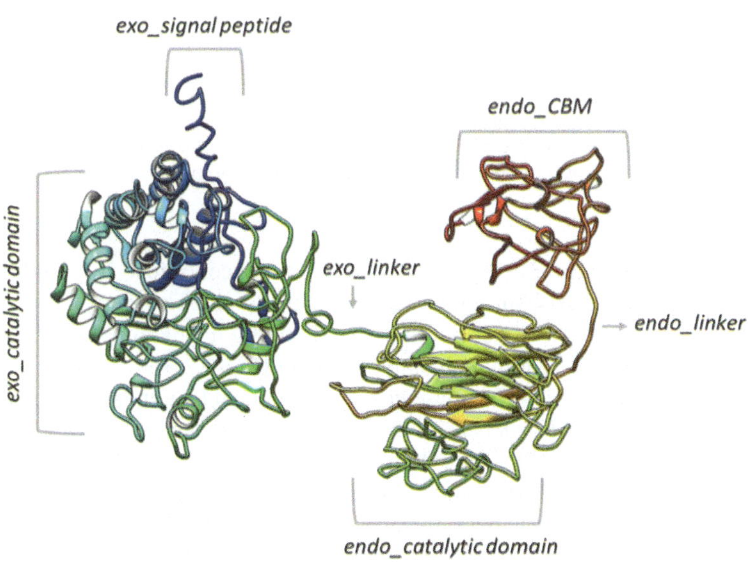

Fig. 12.1 Hybrid exo14-endo13 beta-glucanase enzyme model created by in silico end-to-end fusion of domains

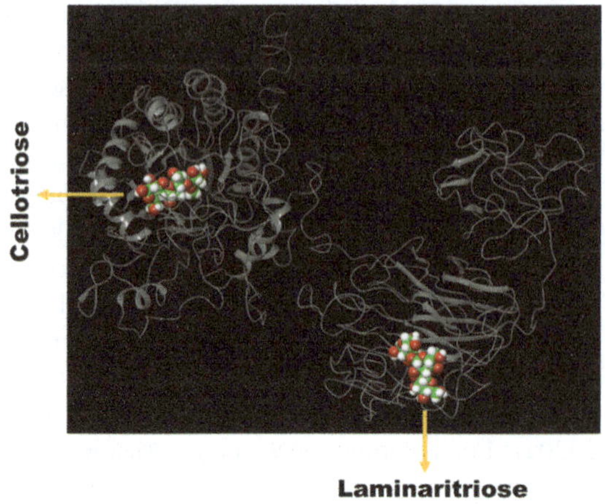

Fig. 12.2 Docking of hybrid exo14-endo13 beta-glucanase enzyme with substrates Cellotriose and Laminaritriose

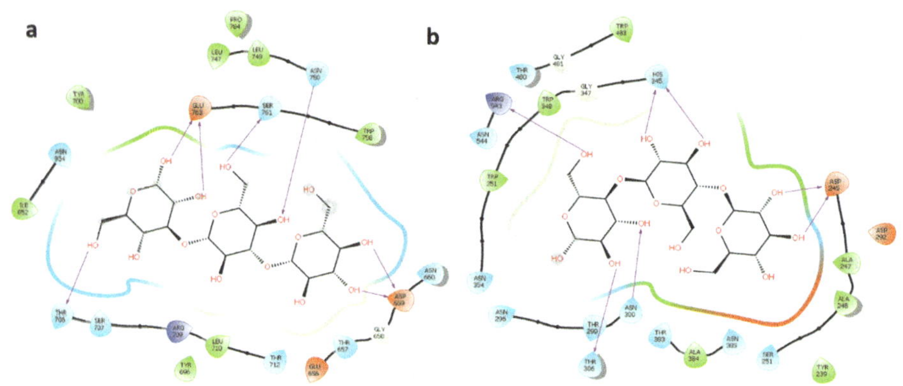

Fig. 12.3 Substrate interaction diagram of catalytic domains of hybrid exo14-endo13 beta-glucanase. (**a**) Interaction of Cellotriose with the catalytic domain of exo-β-1,4-glucanase (exo14). (**b**) Interaction of Laminaritriose with the catalytic domain of endo-β-1,3-glucanase (endo13)

Laminaritriose (for endo13) (Fig. 12.2). The substrate cellotriose is bound with the catalytic domain of exo14 through five hydrogen bonds. The catalytic domain of endo13 is actively involved in seven hydrogen bonding interactions for grasping the laminaritriose molecule within the catalytic pocket (Fig. 12.3).

A hybrid (1–3,1–4)-beta-glucanase enzyme generated by combining the *Bacillus amyloliquefaciens* and *B. macerans* (1–3,1–4)-beta-glucanase genes through an extension overlapping PCR. The hybrid (1–3,1–4)-beta-glucanase contains an N-terminal sequence derived from *B. amyloliquefaciens* and a C-terminal segment from *B. macerans* (1–3,1–4)-beta-glucanase. The mature hybrid enzyme exhibited significant thermostability, stability in acidic pH and 1.5 to 3-fold higher activities than parental enzymes. It was an example of the exchange of homologous gene segments. The hybrid (1–3,1–4)-beta-glucanase improved the characteristics of old versions of (1–3,1–4)-beta-glucanases (Olsen et al. 1991). Wei-Chun Liu et al. 2012, successfully applied a structural-based domain fusion approach to creating bifunctional hybrid beta-glucanases with dual substrate-binding efficiency. The enzyme combined the novel properties of *Fibrobacter succinogenes* 1,3–1,4 beta-glucanase (T*Fs*$_{W203F}$) and hyperthermophilic *Thermotoga maritima* 1,3 beta-glucanase (*Tm*Lam). The obtained hybrid enzymes showed increased specific activities and turnover rates to particular substrates such as lichenan and laminarin. Fusion of endoglucanase (Cel5A) and endoxylanase (XylT) produced a bifunctional enzyme with improved catalytic activities and stabilities in varying ranges of temperatures. The hybrid enzyme exhibited increasing specific activity towards the degradation of substrates beta-glucan and beechwood xylan (Rizk et al. 2015). A divergence from the end-to-end fusion, few studies reveal the application of insertional fusion to produce multifunctional hybrid proteins. A bifunctional hybrid enzyme of laccase and beta-1,3–1,4-glucanase created by an insertion fusion of the genes bglS (beta-1,3–1,4-glucanase) and cotA (laccase) genes from *Bacillus subtilis* showed better hydrolysis of cell wall beta-glucans and also showed the oxidation of aromatic compounds (Furtado et al. 2013).

Studies of hybrid enzymes or the ability to engineer the properties of enzymes will increase the use of enzymes in biotechnological industries. By imitating the natural enzymes, it will undertake a significant role in the scenario. Hybrid enzymes have been efficaciously created through the engineering of enzymatic and non-enzymatic properties. Current advancements in biotechnological approaches in seamless cloning and DNA shuffling, coupled with developing selection strategies will undoubtedly demonstrate innovative avenues in hybrid enzyme technology.

References

Amin SR, Erdin S, Ward RM, Lua RC, Lichtarge O (2013) Prediction and experimental validation of enzyme substrate specificity in protein structures. Proc Natl Acad Sci U S A 110(45):E4195–E4202

Balabanova L, Golotin V, Kovalchuk S, Bulgakov A, Likhatskaya G, Son O, Rasskazov VA (2014) Novel bifunctional hybrid with marine bacterium alkaline phosphatase and far eastern holothurian mannan-binding lectin activities. PLoS One 9(11):e112729

Basit N, Wechsler H (2011) Prediction of enzyme mutant activity using computational mutagenesis and incremental transduction. Adv Bioinforma 2011:958129

Béguin P (1999) Hybrid enzymes. Curr Opin Biotechnol 10(4):336–340

Corey MJ, Corey E (1996) On the failure of de novo-designed peptides as biocatalysts. Proc Natl Acad Sci U S A 93(21):11428–11434

Edison LK, Dan VM, Reji SR, Pradeep NS (2020a) A strategic production improvement of Streptomyces Beta glucanase enzymes with aid of codon optimization and heterologous expression. Biosci Biotechnol Res Asia 17(3):587–599

Edison LK, Pradeep NS (2020) Functional screening of β-glucanase producing Actinomycetes strains from Western Ghats ecosystems of Kerala. India Indian J Appl Pure Biogr 35(2): 255–270

Edison LK, Reji SR, Pradeep NS (2020b) In silico perceptions in structural elucidation of Exo-β-1,4-glucanase (exo14) and endo-β-1,3-glucanase (endo13) from *Streptomyces* spp. J Appl Biol Biotechnol 8(6):67–81
Furtado GP, Ribeiro LF, Lourenzoni MR, Ward RJ (2013) A designed bifunctional laccase/b-1,3–1,4-glucanase enzyme shows synergistic sugar release from milled sugarcane bagasse. Protein Eng Des Sel 26(1):15–23
Golotin V, Balabanova L, Likhatskaya G, Rasskazov V (2015) Recombinant production and characterization of a highly active alkaline phosphatase from marine bacterium Cobetia marina. Mar Biotechnol (NY) 17(2):130–143
Hecht MH (1996) Strategies for the design of novel proteins. In: Carey PR (ed) Protein engineering and design. Academic Press, pp 1–50
Hedstrom L, Szilagyi L, Rutter WJ (1992) Converting trypsin to chymotrypsin: the role of surface loops. Science 255(5049):1249–1253
Kim SB, Izumi H (2014) Functional artificial luciferases as an optical readout for bioassays. Biochem Biophys Res Commun 448(4):418–423
Kim SC, Pósfai G, Szybalski W (1991) A novel gene-fusing vector: construction of a 5′-GGmCC-specific chimeric methyltransferase, M.BspRI/M.BsuRI. Gene 100:45–50
Koga N, Tatsumi-Koga R, Liu G, Xiao R, Acton TB, Montelione GT, Baker D (2012) Principles for designing ideal protein structures. Nature 8(491):222–227
Kohrer C, Rajbhandary UL (2009) Protein engineering. In: Gross HJ (ed) Nucleic acids and molecular biology. Springer-Verlag, Berlin Heidelberg
Liu W-C, Lin Y-S, Jeng W-Y, Chen J-H, Wang AH-J, Shyur L-F (2012) Engineering of dual-functional hybrid glucanases. Protein Eng Des Sel 25(11):771–780
Meunier B (2008) Hybrid molecules with a dual mode of action: dream or reality? Acc Chem Res 41(1):69–77
Nixon AE, Ostermeier M, Benkovic SJ (1998) Hybrid enzymes: Manipulating enzyme design. Trends Biotechnol 16(6):258–264
Olsen O, Borriss R, Simon O, Thomsen KK (1991) Hybrid bacillus (1-3,1-4)-beta-glucanases: engineering thermostable enzymes by construction of hybrid genes. Mol Gen Genet 225(2): 177–185
Otsuka R, Imai S, Murata T, Nomura Y, Okamoto M, Tsumori H, Kakuta E, Hanada N, Momoi Y (2015) Application of chimeric glucanase comprising mutanase and dextranase for prevention of dental biofilm formation. Microbiol Immunol 59(1):28–36
Palmer AE, Qin Y, Park JG, McCombs JE (2011) Design and application of genetically encoded biosensors. Trends Biotechnol 29(3):144–152
Pronk S, Páll S, Schulz R, Larsson P, Bjelkmar P, Apostolov R, Shirts MR, Smith JC, Kasson PM et al (2013) GROMACS 4.5: a high-throughput and highly parallel open-source molecular simulation toolkit. Bioinformatics 29(7):845–854
Reddy Chichili VP, Kumar V, Sivaraman J (2013) Linkers in the structural biology of protein–protein interactions. Protein Sci 22(2):153–167
Reddy VA, Venu K, Rao DE, Rao KV, Reddy VD (2009) Chimeric gene construct coding for bi-functional enzyme endowed with endoglucanase and phytase activities. Arch Microbiol 191(2):171–175
Rizk M, Elleuche S, Antranikian G (2015) Generating bifunctional fusion enzymes composed of heatactive endoglucanase (Cel5A) and endoxylanase (XylT). Biotechnol Lett 37(1):139–145
Schneider WP, Nichols BP, Yanofsky C (1981) Procedure for production of hybrid genes and proteins and its use in assessing significance of amino acid difference in homologous tryptophan synthetase alpha polypeptides. Proc Natl Acad Sci U S A 78(4):2169–2173
Singh A, Hayashi K (1995) Construction of chimeric β-glucosidases with improved enzymatic properties. J Biol Chem 270(37):21928–21933
Singhania RR, Patel AK, Pandey A (2010) The industrial production of enzymes. Ind Biotechnol:207–225

Sinha D, Bakhshi MR, Vora RK, Kirby EP, Budzynski AZ (1996) Synthetic fusion protein design and applications. Anal Biochem 238(2):205–208
Stemmer WPC (1994) DNA shuffling by random fragmentation and reassembly: in vitro recombination for molecular evolution. Proc Natl Acad Sci U S A 91(22):10747–10751
Turanli-Yildiz B, Alkim C, Cakar ZP (2012) Protein engineering methods and applications. Protein engineering. In: Kaumaya P, ed. Tech. Available from http://www.intechopen.com/books/protein-engineering/protein-engineering-methods-and-applications
Warrens AN, Jones MD, Lechler RI (1997) Splicing by overlap extension by PCR using asymmetric amplification: an improved technique for the generation of hybrid proteins of immunological interest. Gene 186(1):29–35

The manufacturer's authorised representative in the EU is Springer Nature Customer Service Centre GmbH, Europaplatz 3, 69115 Heidelberg, Germany. If you have any concerns regarding our products, please contact ProductSafety@springernature.com

Printed and bound by CPI Group (UK) Ltd, Croydon, CR0 4YY
30/09/2026
02225228-0006